Advances in Asian Human-Environmental Research

Aims and Scope

The series aims at fostering the discussion on the complex relationships between physical landscapes, natural resources, and their modification by human land use in various environments of Asia. It is widely acknowledged that human-environment interactions become increasingly important in area studies and development research, taking into account regional differences as well as bio-physical, socioeconomic and cultural particularities.

The book series seeks to explore theoretic and conceptual reflection on dynamic human-environment systems applying advanced methodology and innovative research perspectives. The main themes of the series cover urban and rural landscapes in Asia. Examples include topics such as land and forest degradation, glaciers in Asia, mountain environments, dams in Asia, medical geography, vulnerability and mitigation strategies, natural hazards and risk management concepts, environmental change, impacts studies and consequences for local communities. The relevant themes of the series are mainly focused on geographical research perspectives of area studies, however there is scope for interdisciplinary contributions.

More information about this series at http://www.springer.com/series/8560

Padmini Pani

Land Degradation and Socio-Economic Development

A Field-based Perspective

Springer

Padmini Pani
Centre for Study of Regional Development, School of Social Sciences
Jawaharlal Nehru University
New Delhi, India

ISSN 1879-7180 ISSN 1879-7199 (electronic)
Advances in Asian Human-Environmental Research
ISBN 978-3-030-42073-4 ISBN 978-3-030-42074-1 (eBook)
https://doi.org/10.1007/978-3-030-42074-1

Cover illustration: Nomads near Nanga Parbat, 1995. Copyright © Marcus Nüsser (used with permission)

This Springer imprint is published by the registered company Springer Nature Switzerland AG.
The registered company address is: Gewerbestrasse 11, 6330 Cham, Switzerland

To my parents,
Arun Kumar Pani
and
Bela Pani

Preface

Land degradation is a complex problem affecting a large section of poor people in several countries. It is a process that is affected both by natural and human-induced factors, which do not operate in isolation. Land is the primary source of a large section of the population in developing countries like India. Because of the high share of people dependent on agriculture and high population density, soil health is an essential aspect of overall food and nutrition security in developing countries. Addressing land degradation is among the crucial steps for achieving Sustainable Development Goals of food and nutrition security. While there are ongoing efforts to combat desertification and land degradation at the global and national levels, land degradation in the context of growing land hunger has emerged as a significant concern for sustainable development. With rapid urbanisation, diversion of land from agricultural to non-agricultural uses, and within agriculture, from food to non-food crops, sustainable land-use practices have acquired a new significance. However, for effective interventions at the ground, a bottom-up approach is needed. Peoples' perspectives are critical for understanding the problems associated with land degradation as well as for developing effective solutions.

This book is an attempt to understand the interface between land degradation and socio-economic development in a specific regional context. The Chambal region, which is the focus of the study, is well known for its badlands and social problems associated with lack of development, conflicts, violence and insecurity. This book seeks to unravel the linkages between the environmental processes associated with land degradation and the economic and social processes associated with economic development.

The Chambal Badlands, despite being located in close proximity of major cities like Agra and Gwalior, is among the relatively less developed regions of India. In the popular imagination, the region is often associated with the prolonged presence of bandits and high rates of crime. The rugged terrain and undulating topography of the region make it highly inaccessible. Because of the inaccessibility and problems of communication, this region is also among the less researched areas of India. Furthermore, the available research on this region, particularly from the perspective of land degradation, appear to be highly fragmented. The core objective of this book is to connect the natural and socio-economic processes associated with land degradation. Global research on land degradation and desertification has brought out the inherent complexities of these interactions between the natural and social processes. Land degradation not only is partly caused or aggravated by the anthropogenic factors but also affects the socio-economic processes. The natural resource base of a region offers opportunities for economic development, and environmental degradation, in general, and land degradation, in particular, affect the prospects of future growth. Also, in the analysis of the social and economic causes and outcomes of land degradation, human society cannot be taken as a homogeneous entity. There are social groups which have unequal resources to natural resources, such as land and forests.

Furthermore, marginalised social groups often have less access to government-sponsored programmes. Gender division of labour and unequal access to resources by women and children usually mean that they are affected differently by environmental degradation. At times, the response of the relatively better-off and marginalised sections to a similar environmental crisis could be very different from each other. In this multidisciplinary study, an attempt has been made to understand the implications of land degradation, by looking at the differences in the levels of development between villages affected by land degradation and those not affected by it. Furthermore, the changes in the livelihoods of those affected by land degradation have also been studied.

The study emphasises the need to design sustainable development strategies, considering the fragilities of the local ecology. In a sense, Chambal Badlands has some unique ecological properties, and any kind of development interventions should be based on a nuanced understanding of the relationship between economic and ecological processes. Though based on the study of specific badland region in India, the findings of this research have broader significance in understanding the processes of human-nature interactions.

New Delhi, India Padmini Pani

Acknowledgements

A study on the causes and implications of land degradation in one of the less developed yet less well-researched regions of India needs no justification. The Chambal region provides a formidable challenge to any researcher not only because of its difficult topography but also because of the lack of systematic investigations into the nature and causes of its backwardness. This study is a modest attempt to fill up this research gap, and it would not have been completed without the help of a number of institutions and individuals. First of all, I am grateful to the Indian Council of Social Science Research, New Delhi, for sponsoring this research. I am thankful to the Jawaharlal Nehru University, New Delhi, and the Centre for the Study of Regional Development, JNU, where the study was hosted.

I would like to thank a number of people for their help, constructive suggestions and critical observations which help in improving the work. I am grateful to the villagers who patiently answered my questions during the various rounds of field survey and helped me to understand the landscape of the area. I am indebted to the people of the Chambal region. Without their generosity, it would not have been possible to get access to the remote badlands and closely knitted villages, and my understanding about the area and its geomorphology as a whole would have been incomplete. I am thankful to all the research assistants, field assistants and field guides for their help in conducting this research. Particularly, I would like to thank Professor Deepak K. Mishra, CSRD, JNU, for his valuable suggestions and criticisms. Especially after the field visits, his insightful discussions always helped me to integrate the physical phenomena and social processes. I am also thankful to Professor S. N. Mohapatra, Head, School of Earth Science, Jiwaji University, Gwalior, for his help to explore a larger perspective of the study region and its geological dynamics. I am also grateful to my research scholars at CSRD, JNU, many of whom have been conducting research on broadly similar concerns. Support from administrators, government officers and officials is gratefully acknowledged. Lastly, I would like to express my gratitude to my ailing father for his patience and my mother for her empathy which provided me with the strength to complete this work. Last but not the least, I thank my son Rudraksh who not only had to endure my long absences during the field survey but also was a cheerful companion at different stages of the work.

The findings of the project have been presented in a number of national and international conferences and have also been published in a few national and international journals. I am thankful for the suggestions and criticisms that I received from the participants and the referees of the journals. Needless to add, I am solely responsible for the shortcomings of the study.

Padmini Pani

Contents

About the Author

Padmini Pani is Associate Professor of Geography at the Centre for the Study of Regional Development, Jawaharlal Nehru University, New Delhi. She has extensively worked in the areas of land degradation, environmental geography and fluvial geomorphology. She has coedited *Geoinformatics for Natural Resource Management* (Nova Science, 2009). She has been the recipient of several prestigious research awards and has carried out collaborative research at universities in the UK, Germany and Canada.

List of Figures

List of Plates

List of Tables

Chapter 1
Introduction

Abstract The first chapter is an introduction of the theme and a global view which dealt with all aspects of land degradation, related to environmental degradation and development, links between land degradation and development; meaning, definition and the types of land degradation; causes, processes and measurements of land degradation; the land degradation scenario in India; and the linkage between land degradation and rural development. The objectives, database and adopted methodology for the study are also placed in this section.

Keywords Land degradation · Gully and ravine erosion · World scenario of land degradation · Meaning and definition of land degradation · Methods of land degradation analysis · Remote sensing and geographic information system (GIS) · Land degradation in India

1.1 Introduction

Land degradation is among the key environmental challenges before mankind. Food is a basic requirement of human society, and to a great extent, the ability to provide adequate food to each and every member of the society depends on soil health. The Sustainable Development Goals cannot be meaningfully achieved without taking care of the problem of land degradation. In many developing and less developed countries, land degradation creates barriers for sustainable development.

1.1.1 Environmental Degradation and Development

"Degradation" is, by description, a process of change over time (Scherr and Yadav 1996). Environmental degradation is the deterioration of the environmental resources through reduction of some important life support resources like air, water

P. Pani, *Land Degradation and Socio-Economic Development*, Advances in Asian Human-Environmental Research,
https://doi.org/10.1007/978-3-030-42074-1_1

and soil, etc. Land degradation and loss of biodiversity create an immense pressure for the world's ecosystems, and their capacity to support the important services is decreasing at the point of time when the need for these services is even more (MEA 2005). The concept of environmental quality and ecological risk has become a core concept that connects to socio-economic change and political dynamism (Salvati et al. 2015). Recent approaches are human and sustainable development. These two concepts basically argue to consider development to a broader perspective rather than a mere concept of economic growth. Much have been written and discussed about these issues ever since their evolution. These approaches have worked well to redefine development to a newer concept but need further revision in addressing the issues of sustainability of development (Ross 2009). Defined by Brundtland Commission 1987 and expanded by the Earth Summit 1992, sustainable development considers three pillars of development: economic, social and environmental growth (Soubbotina 2004). Each pillar contains a number of development indicators that must be satisfied by the countries. Despite the tremendous development achievements in the new millennium, there are contradictions abound. There are debates on the cost of development against cost of environment. Are countries able to attain a balance between development and environmental sustainability? Who are paying for the human cost of development? What should be basis for achieving sustainability in the development? There are number of questions like this that the development discourse has to propose further. The relationship between development and environment is always contending. There is no straightforward relationship of environmental impact of development. Human impact on environmental degradation is imposed by developmental activities that demand careful analysis (Pokharel 2015). Thus, defining development is tricky and contextual. But it is necessary to define and act on. There are enough theoretical writings on putting human face into development and ensure the sustainability of development outcomes. But, the results are yet to confirm the practices. There is a trending gap in ecological balances. Since the principle of human development is to enlarge choices of people, such choices are not uniform throughout the world (Pokharel 2015). The destruction of ecosystems, natural habitat for all living beings, puts a barrier to the enlargement of choices of individuals.

Land degradation is ascertained to be one of the severe global environmental challenges (Eswaran et al. 2001; Lal 2001; Scherr and Yadav 1996). It has numerous economic, social, ecological and environmental implications. Land degradation is an important geomorphic process in many parts of the world in a range of landscapes and its causal determinants. Natural causes which incorporate climate change, catastrophic storms and isostatic rebound, tectonic uplift or base level lowering, removal of vegetation and deforestation, overgrazing, population pressure, unsuitable agricultural practices, institutional settings and public policy are among the significant anthropogenic factors causing land degradation (Pani and Carling 2013). The relations between social and natural processes impacting on land degradation are usually complex and locale specific (Blaikie 1985). Land degradation is having impact on billions of people around the world and may lead to the overexploitation of soil resources, decline of groundwater resources, loss of ecosystem productivity, shifts in vegetation composition and loss of rural livelihoods (Schröder et al. 2016). The impact of land degradation and the degradation of environment are related to each other, and a change of one component of it changes the balance of environment and creates problem or threat to the land and its sustainability.

1.1.2 Land Degradation: A Global Overview

Land is one of the most valuable natural resources which ensure food security for all sections of people. Many of the developing and less developed countries are heavily dependent on land productivity as most of the people's livelihood depends on agriculture or land-related resources. Land degradation is growing steadily and spreading in many parts of the world. More than 20% of all cultivated areas, 30% of forests cover and 10% of pasture are suffering degradation (Bai et al. 2008). It has been observed that irrespective of difference of climatic regions, the millions of hectares of land each year are being degraded. It has been assessed that more than hundred countries, influencing over 33% of the earth's land surface and 2.6 billion people, are getting affected by land degradation and desertification (Adams and Eswaran 2000). As an environmental issue having implications throughout the world, land degradation has been a major concern at the United Nations Convention to Combat Desertification, the Convention on Biodiversity, the Kyoto protocol on global climate change and the millennium development goal (United Nations Conference on Environment and Development, UNCED 1992; United Nations Environment programme, UNEP 2008). The occurrence is most noticeable in the drylands, which cover more than 40% of the earth's surface (Dobie 2001). At present approximately 73% of grassland in dryland areas are degraded, along with 47% of marginal rainfed agricultural land and a significant percentage of irrigated agricultural land (United Nation Convention to Combat, UNCCD Agenda 21, 1992). About 20% of the pastures and rangelands have also been damaged by overgrazing (Food and Agricultural Organisation, FAO 1996). Two of the most important studies on land degradation designed for purposes of international comparison are the Global Land Assessment of Soil Degradation (GLASOD) mapping exercise by Oldeman et al. (1991) and the comparative study of drylands by Dregne and Chou (1992). As per the GLASOD estimation nearly 8.7 billion hectares of agricultural land, pasture, forest and woodland, since mid-century about 2 billion hectares (22.5%) have been degraded. Around 3.5% has been degraded irreversibly, and without a costly engineering measure, this percentage of land is not possible to be retrieved. About 10% has been moderately degraded, and bringing back its health is possible through on-farm investments only. The rest of nearly 9% is degraded but possibly reversible through good land husbandry practices. Another study of GLASOD stated that in Asia approximately 27% of total agricultural land, permanent pasture and forest and woodland areas are affected by some form of soil degradation. Interestingly the affected lands are mostly in dry regions which are more than 50%.

1.1.3 Land Degradation and Development

Many scientists and social scientists believe that land degradation is a potential threat in long-term food security globally (Pimentel et al. 1995; Brown and Kane 1995); however some scientists reserve their views about it (Crosson 1994). Land degradation façades a significant problem to agricultural growth and reducing poverty in India. It is officially calculated that around 44% of the land in India is degraded (Mythili and Goedecke 2016). A steady acceleration in environmental pollution and degradation of natural resource for the past few decades have been identified as measures of developmental challenge globally included in the first global landmark event – the Human Environment Conference in Stockholm in 1972 (World Bank 2010). Spatiotemporal studies on land degradation can provide insights on spatial distribution of vulnerable areas that can enhance ecological understandings and policy making for regional development (Contador et al. 2009; Salvati et al. 2011). GLADA (Global Assessment of Land Degradation and Improvement) has identified land degradation as long-term decline in the ecosystem function in a broader perspective, which can be measured in terms of NPP (net primary productivity) that can help visualise development. Consequently, the area with degraded land has low level of socio-economic development (Bai et al. 2008; ISRIC & FAO 2010). It is well established that the sustainable development has anticipated for a consistent effort to bring together growth with environmental quality (Salvati et al. 2015). From a broader perspective, it can be suggested that land degradation is the mismatch between land quality and land use that is affecting development.

The relationship of land degradation and development is important but is yet to be explored enough. The literatures suggest that development which is measured in terms of economic growth is usually accompanied by environmental degradation. Some developing countries have been facing threats because of land degradation up to 1–17% of the gross national product, and the rate of loss has reached 10% in some of tropical countries (Feng et al. 2005). Further, land degradation is a biophysical process driven by political and socio-economic causes that affects developmental initiatives (Eswaran et al. 2001). Land degradation is influenced both by natural and anthropogenic factors, and its consequences are strongly related to people as it lessens the services provided by the territorial ecosystem that affects development (Gerber et al. 2014).

The link between degradation and development is another important issue which is yet to be explored fully. This relationship is widely known as the 'Environmental Kuznets Curve' (EKC), which is often described through a bell-shaped relationship between environmental degradation and per capita income. However, based on the experience of Indian agriculture, it is propositioned that agricultural sustainability moves cyclically as the level of development increases. According to this view, along with the level of development, sustainability increases (degradation or unsustainability declines) in the case of fragile regions (lie on the rising portion of the 'bell'-shaped curve), whereas well-endowed regions experience stagnation or decline in sustainability (lie on the top or the falling portion of the curve). When viewed from unsustainability or degradation point (EKC style), the relationship reflects a 'U'-shaped curve rather than a 'bell'-shaped one. However, it may be noted that this is purely a theoretical proposition and hence calls for empirical exploration. On the whole, it may be concluded that while the poor are definitely the victims of degradation, the evidence does not support the argument that poor are the agents of degradation (Reddy 2003).

Land degradation and sustainable development is critically linked specially in agrarian economies like India. Often small land holding farmers face the problem of low land productivity where land is degraded. This is attributed mainly to the population, agriculture and environment nexus. It was observed that increased population density without technological progress has led to reduced fallow periods, soil degradation and deforestation (UNSO 1993). On the contrary, suitable institutional actions could cease or reduce degradation even under the circumstances of snowballing population pressure (Reddy 2003).

The extent and costs of degradation are significant in India. Being an agrarian economy, any aggravation in the situation would make threats in food security of the nation. The adverse impacts of degradation are though limited to certain areas. As far as mitigating measures are concerned, the ARPU at Ahmedabad has come out the necessary policy interventions at the regional level (Wadia 1996). These involvements are based on detailed regional studies. The various components of land degradation such as salinity, waterlogging and wind and water erosion are closely associated with water and forest resources. Judicious management of all these components is vital for achieving overall sustainable development. Population pressure does not appear to apply any excessive pressure as far as land degradation is concerned. Interestingly the major interventions in all the problem regions relate to water and land management. It is common knowledge now that these resources are on the decline, quantitatively as well as qualitatively. This could be the indication of more serious environmental problems that might affect future generations. Poverty-stricken populations residing in fragile environmental resource areas are the most vulnerable of the process regardless of the fact that they do not contribute to the process. Ecological management of natural resources demands for proper market and institutional mechanisms. Markets for natural resources either do not exist or are inaccurate.

1.1.4 Land Degradation: Meaning and Definitions

In general, when a land lost its biological productivity due to the biophysical processes, the land is considered as a degraded land. Defining the exact meaning or term, "land degradation" is complex and tough as well mainly at present circumstance. This is an extensively used term to imagine diverse, overarching definitions. The term 'land' is a 'terrestrial bio-productive system that comprises soil, vegetation, other biota, and the ecological and hydrological processes that operate within the system', and its "degradation" is a 'reduction or loss ... of the biological ...

productivity, ... resulting from land uses, ... or combination of ... processes, such as... soil erosion ... deterioration of ... properties of soil ... and long-term loss of natural vegetation'. Applying the more recent conceptual framework of the Millennium Ecosystem Assessment (MEA[1]2005) rather using this long definition better to define it as a terrestrial ecosystem", and "land degradation" – to "reduction or loss of ecosystem services, notably the primary production service" (Safriel and Adeel 2005). As per Blaikie and Brookfield (2015), 'land degradation' concept got its viability only in the social context of the usefulness obtained from use of ecosystems by the human beings. Therefore, most of the cases, land degradation 'is frequently derived directly by evaluating the soil, the vegetation cover and its primary productivity. Commonly in an agrarian society of a developing and under developed countries the land is one the primary means of production' (Grepperud 1997; Nkonya et al. 2008). Land resources are generally controlled by the many other resources such as water, forest, etc. The poor management of any of them would affect the land quality. Apart from its agricultural practice, mining and other human activities are directly having impact or linkage with the land quality and land. Land is one of the severely affected natural resources in India. But as per NRSC the degraded term should be used for those lands where due to the natural causes or mismanagement and lack of proper irrigation facility when a land is no longer productive enough, we considered the degraded land term, but the wasteland is land in other side wasteland is the result of imposed inherent or imposed disabilities such as location, environment and physical and chemical properties of a soil. It is better to define land degradation as per the millennium ecosystem assessment report as the long-term loss of ecosystem services that affects the livelihood and food security of billions of people.

As a resource, study of land not only covers the soil resource but also includes the water resource, vegetation coverage, landscape characteristics and even the micro-climatic components of the terrestrial ecosystem (Scherr and Yadav 1996). Land degradation is a process in which biological productivity of land is reduced that has varied social, economic and ecological consequences (Pani and Carling 2013). It relates to long-term changes in ecosystem functions. It causes alteration in the physical and chemical qualities of soils that can be noticeable via loss of soil nutrients, fall in productivity of land, loss of biodiversity and downfall of economic viability (Feng et al. 2005). Land degradation as a process has been the matter of discussion at the global level, and there are two distinct thoughts regarding the forecast, severity and effect of land degradation (Eswaran et al. 2001). One school considers it as a severe global risk causing adverse effect on biomass productivity and environment quality that scholars like ecologists, soil scientists and agronomists cover. The second school emphasises land degradation as vital issue, and land managers (especially farmers) have dependency on land productivity and profits; it is their interest to maintain land (Eswaran et al. 2001).

[1]MEA is Millennium Ecosystem Assessment. For details see MEA (2005).

In the Indian context, it is very common that the wasteland and degraded land are used as synonyms. However, the description of wasteland 'degraded land which can be brought under vegetative cover with reasonable effort, and which is currently under-utilized and land which is deteriorating for lack of appropriate water and soil management or on account of natural causes. Wastelands can result from inherent/imposed disabilities such as by location, environment, chemical and physical properties of the soil or financial or management constraints' (National Remote Sensing agency, NRSA 1995). The nomenclature 'wasteland' is misleading as it suggests that these lands do not have any economic or ecological value. Therefore, we prefer the term 'degraded land' throughout the paper with due regard to their economic and ecological contributions, which are more familiar to local communities than to the policy makers. NRSA does not provide data under specific headings of wind and water erosion. These are covered under gullied/ravenous land and land with or without scrub. In the case of other estimates, data are given under water and wind erosion, which account for a substantial portion of degraded lands. For the purpose of the present study, we tried to use these various sources appropriately. Besides, there are other categories of land, degraded or non-degraded, which include permanent or other fallows and cultivable waste. These lands are termed as underutilised lands. These lands are not considered separately, as the degraded part of these lands is already accounted.

1.1.5 Land Degradation: Types

Land degradation is associated with the reduced productive capability of a land, either in the short or in the long term (UNEP 1992). Water resources, forest and air, when negatively impacted by the human activities, might result in various types of land degradation. Due to the anthropogenic impact, land degradation, specifically soil erosion, caused by water and wind erosion, results in the deterioration of soil properties (viz. physical, chemical and biological properties). The soil resource degradation includes soil erosion by water and wind. Soil degradation leads to the deterioration of physical, chemical and biological properties of soil, waterlogging and the build-up of toxicities, particularly salts, in the soil. Subsequently soil productivity is closely associated with water availability; declining of the groundwater table is also prominent. There are several ways and processes adopted to classify the types of land degradation. Broadly, these can be classified into six classes: water erosion, wind erosion, soil fertility deterioration, salinisation, waterlogging and deceasing of the water table.

Land degradation has a wide range that covers primarily five broad components that are physical soil management, soil water management, soil nutrient and organic matter management, soil biology management and vegetation management. The soil management physically mainly initiated of crusting, compaction, sealing, wind erosion, water erosion, devegetation, over tillage, etc. The soil water management created impeded drainage, waterlogging, reduced water holding capacity and infiltration and salinisation. The other causative factors for the alkalisations, acidifications, nutrient leaching, removal of organic matter, vegetation burning residues and nutrient depletions of the soils are soil nutrient and organic matter management. The soil biology management affects the decline in vegetative cover and biodiversity and affects the species composition with lack of availability of valued species (Scherr and Yadav 1996.).

According to the NRSA definition in 1995, degraded land is classified into 12 categories, which include forest-based degradation, sandy, hilly and rocky and snow-covered regions. However, in the context of land (non-forest) degradation, eight classifications are important. These include (i) gullied and ravenous land, (ii) upland with or without scrub, (iii) waterlogged and marshy land, (iv) land affected by salinity and alkalinity, (v) shifting cultivation area, (vi) degraded pasture land and grazing lands, (vii) degraded land under plantation crops and (viii) mining and industrial wastelands (NRSA 1995). Land degradation has been reclassified again into 23 categories which dealt with the degradation types in more details (Table 1.1) using different temporal satellite imageries like Resourcesat-1 LISS-III imagery for years 2005–2006 and 2008–2009 belonging to all three major cropping seasons of India

Table 1.1 Regrouping of 23-fold wasteland classification into 12-fold classification

S. No	Wasteland class		Regrouped class
1	Gullied and/or ravinous land (Medium)	1	Gullied and ravinous land
2	Gullied and/or ravinous land (Deep)		
3	Land with dense scrub	2	Scrub land
4	Land with open scrub		
5	Waterlogged and marshy land (Permanent)	3	Waterlogged area
6	Waterlogged and marshy land (Seasonal)		
7	Land affected by salinity/alkalinity (Moderate)	4	Salt affected area
8	Land affected by salinity/alkalinity (Strong)		
9	Shifting cultivation – current jhum	5	Shifting cultivation
10	Shifting cultivation – abandoned jhum		
11	Underutilised/degraded forest (Scrub domin)	6	Degraded forest
12	Underutilised/degraded forest (Agriculture)		
13	Degraded pastures/grazing land	7	Degraded pastures/grazing land
14	Degraded land under plantation crop	8	Degraded land under plantation crop
15	Sands – riverine	9	Sandy area
16	Sands – coastal		
17	Sands – desertic		
18	Sands – semi stab.; stab >40 m		
19	Sands – semi stab.; stab 15–40 m		
20	Mining wastelands	10	Mining/industrial wastelands
21	Industrial wastelands		
22	Barren rocky area	11	Barren rocky area
23	Snow covered/glacial area	12	Snow covered/glacial area

Source: NRSA (2011)

(*kharif, rabi and zaid*) with reference to Survey of India topographic maps on 1:50000 for scale. Published land use/land cover maps, reports atlases have also been used to prepare the above said categories with limited ground truth verifications.

1.1.6 Land Degradation: Causes and Processes

The GLASOD assessment is based on each type of the degradation and its cases. In this evaluation, only four causes were identified, these area stated as deforestation and of natural vegetation removal, overgrazing, agricultural activities and overexploitation of vegetation for domestic use. However, irrigation as a cause of land degradation has not been considered.

As per the FAO, the causes of land degradation can be divided into (a) natural hazards, (b) direct causes and (c) underlying causes. The presence of high degradation caused the physical environment and processes under natural hazards, like steep slopes accelerate water erosion. Land degradation incorporates the whole environment. Nevertheless, researchers take into account individual factors concerning soils, water resources including surface and ground, forests, grasslands, croplands and biodiversity (FAO 2005). The unsuitable land use and inappropriate land management practices come under the direct causes. For example, cultivation of steep slopes without measures for soil conservation results in degradation. There are various underlying causes behind the inappropriate types of land use and management. For example, the cultivation on slopes could be because the landless poor need food and conservation measures are not adopted because of insecure property rights (FAO 2005). Often, land degradation is caused by inappropriate land management such as deforestation, overgrazing, monoculture, salinisation, over-use and unscientific use of fertilisers, pesticides and other chemicals, unsustainable farming practices and soil erosion.[2] The summary of GLASOD's result of assessment of causes along with the insights from other published works is presented in Tables 1.2 and 1.3.

The fact has already been established that the human activities are either the major causes of land degradation or aggravating the degradation. The deterioration in land quality triggered by human actions has been a key global issue since the twentieth century and will continue high on the worldwide agenda in the twenty-first century (Eswaran et al. 2001). The instantaneous causes of land degradation are the result of unsuitable land use practice which leads to soil and vegetation cover loss and affects the other biological diversities and ecosystem structure and functions (Snel and Bot 2003). Climate change impacts, viz. increased temperature, severe droughts and erratic rainfall, on degraded lands are known to be more severe.

The causes of land degradation are several and compound. At times the same causal factor could lead to variations in outcomes with implications in different

[2] Available at (http://www.fao.org/docrep/v4360e/V4360E08.htm#Direct%20causes%20of%20degradatio)

Table 1.2 Causes of degradation as given in the GLASOD assessment

Type of degradation	Percentage area of degradation type caused by			
	Deforestation	Overgrazing	Agricultural activities	Overcutting of vegetation
Water erosion	61	67	2	44
Wind erosion	21	46	1	98
Soil fertility decline	25	0	75	0
Salinisation	34	30	14	87
Waterlogging	0	0	85	33
Lowering of water table	12	22	65	34
All types of degradation	37	46	15	63

Source: GLASOD

contexts, because of its changing interactions with other immediate and underlying causes of land degradation. Further, it is important to state that in many cases the causes and types do not necessarily remain same as it changes over time and space.

1.1.7 Measuring Land Degradation

Measuring land degradation is a very complex exercise. It needs to deal with the scale, nature and severity of land degradation. It can be assessed at the global level, regional level, plot level or farm level. Therefore, the nature of assessment is different in each level. For global level indicators, remote sensing and GIS (geographic information system) are very useful for mapping land degradation, vegetation loss, biodiversity and the change of degradation status. At the regional level, drylands, rangelands, grasslands, forests, deserts, soils and rivers systems can be assessed. Interviews, questionnaire and focus group discussions are some of the useful assessment methods at the ground level. Apart from it the different remote sensing techniques like NDVI, NSDI, etc. are useful to measure the degree of degradation. There are several methods and modelling like soil erosion modelling, RUSLE method and field monitoring that are useful to get a statistics of land degradation regionally. Field monitoring and measurement is useful to verify the models. Vegetation change, land cover, land use and slopes can also be assessed through such models that are also useful to study the impact, causes and risks of soil erosion, productivity, etc. For lands, soils, cropland lands, grasslands, forests, conserved area, deserts, grass land, rivers, etc. that come under the local level, following methods like indicators, questionnaires, interviews, focus groups, stakeholders perceptions, special remote sensing and GIS techniques like MODIS, NDVI, MSDI, and different models, for example, USLE/RUSLE (Universal Soil Loss Equation/Revised Universal Soil Loss Equation), can be used with field monitoring and measurements which would help to test the models in farm level. It is difficult to evaluate the tangible

Table 1.3 Proximate and underlying drivers related to land degradation and their potential cause-effect mechanisms

Drivers	Type	Examples of causality	References
Topography	Immediate causes	Steep slopes are prone to soil erosion	Wischmeier (1976), Voortman et al. (2003)
Land use/land cover change	Natural and anthropogenic	Transformation of vegetation cover, grass land cover to agricultural land cause of top soil loss, soil erosion, soil salinity, deforestation	Gao and Liu (2010)
Soil vulnerability	Proximate and natural	Clay reached soil is more prone to erosion	Sharma (1980), Bonilla and Johnson (2012), Pani (2016, 2017a)
Climate	Proximate and natural	Semiarid and arid are prone to erosion. Low and frequent rainfall, torrential rainfall, salinisation	Safriel and Adeel (2005), Barrow (1991), Pani and Carling (2013)
Pest and diseases	Proximate and natural	Pests and diseases lead to soil acidification, loss of biodiversity, loss of livestock and many other forms of land degradation	Sternberg (2008)
Unsuitable land management	Proximate and anthropogenic	Land clearing, overgrazing, terrace cultivation, pollution and siltation of water sources, soil nutrient loss, mining, etc. are the causes of land degradation, terrace farming, forest clearing	Nkonya et al. (2008, 2011), Pender and Kerr (1998)
Population pressure	Underlying	Population pressure, land pressure due to the population density, intense agriculture due to the high population pressure	Pani and Carling (2013), Pani (2016, 2017a)
Land tenure	Underlying	Insecurities of land, uncertainty of land ownership, lack of proper sustainable land maintenance and management	Kabubo-Mariara (2007) Blaikie (1985), Besley (1995)
Poverty	Underlying	There is a no definite answer, but certainly there is a strong linkage between poverty and land degradation	Blaikie and Brookfield (2015)
Policies	Underlying	Policy certainly accelerates land degradation; it may be internationally or state level. Land levelling is a policy in Madhya Pradesh, one of the gullies- and ravine-affected states, but due to the land levelling policy, there are on-site and off-site impacts	Pani (2016), (2017a, b), Marzolf and Pani (2017)
Infrastructure development	Proximate and anthropogenic	Tractor, earth remover, transports, trucks, tractors,	Pani (2017a), Geist and Lambin (2004)

Source: Adopted and modified after Braun et al. (2013)
Note: The cause and effect shown here are not exhaustive

extent and effect of land degradation. Farmers most of the time hide the impact of degradation by adopting their land to less stressful uses or growing levels of rewarding inputs. There is hardly a one-to-one connection among the extent of degradation and the impact on yields (Scherr and Yadav 1996).

1.1.8 Land Degradation in India

India being an agrarian economy, the land is a primary asset for national development. The impact of land degradation in the respect of productivity was estimated to be 4.5% of GDP in 1992 in India as the cost of land degradation (Tata Energy Research Institute, TERI 1998). Several case studies have stated land degradation in Indian context is regarded in terms of decreasing land use intensity, high input (fertilisers) use (in terms of cost) and declining profits (in terms of productivity) (Reddy 2003). The annual costs of land degradation by changing land use and land cover were estimated in 2009, to amount to 5.4 billion USD, compared to 2001. The highest costs were noted in the states of Orissa, Madhya Pradesh, Kerala, Rajasthan and Andhra Pradesh. In every state, the costs of delay surpass the prices of accomplishment. On average, the costs of action are less than half the costs of delay (41%), and regional differences show much lesser costs of inaction in Punjab and Haryana due to limited change of land use. Grassland degradation leads to a loss of 7.7 million USD in meat and milk production. According to the estimation by NRSA, only 20% of the territory in India, i.e. 65 million ha of land, are considered as wastelands. However, these two estimates do not necessarily contradict each other because the effort is to measure different aspects. As per one study by National Soil Bureau of Soil Survey and Land Use Planning in 1994, it is estimated that up to 187.8 million hectares of land are under soil degradation, and an estimation by TERI (1998) revealed that it has increased – as per the estimate, it was 112 million hectares in 1947 (TERI 1998). Various agencies and organisation provide data regarding land degradation. In context to Indian agencies, the NRSA (National Remote Sensing Agency) has given estimates about five broad categories of degradation, namely, water erosion, wind erosion, waterlogging, salinity and nutrient depletion. NRSA provides information on gullied/ravenous lands and lands with or without scrub, which represent water and wind erosion (Reedy 2003). The estimate of the wasteland and land degradation is given below in figure; the estimation has been done by NRSA in 2010 and 2016. As per recent assessments by the Indian Space Research Organisation, in 2007, 32.07% (105.48 mha) of the total geographical area of the country was disturbed by different types of land degradation. About 10.21% (33.56 mha) of the total degraded area was because of water erosion. State-level approximations from the same study recommend that in Madhya Pradesh, of the total 3.5 mha of degraded land, nearly 37.92% was caused by water erosion (Space

Application Centre (SAC)[3], Indian Space Research Organisation (ISRO)[4] 2007). An assessed 3.98 million ha of ravines appear along the rivers Yamuna, Chambal, Mahi, Sabarmati and others (Ministry of Agriculture 1984). In a ravine area, it seems that the first aerial and satellite-based study for mapping of cropland erosion was carried out by Stephens and Chilar (1981, 1982). Recently remote sensing data has been extensively used to monitor changes in spatial extent of ravines. Pani and Mohapatra (2001) observed an increase of ravine-affected area, and the villages have constantly been shifting from their original locations in response to soil degradation and erosion. The evaluation of different landform units and the implication of neotectonic activity in the area using remote sensing techniques have been carried out by Pani et al. (2005, 2008) who carried out a study of erosion intensity assessment in a part of the lower Chambal valley. Chatterjee et al. (2009) have presented a characterisation of the ravines as a function of their erosion prospective articulated through ravine density and ravine depth. The ravine surface cover was estimated in quantitative terms exploiting the preferential characteristics of side-looking, long-wavelength, coherent SAR signal and precision measurements associated with the SAR (Synthetic Aperture Radar) technique. The study of farmers' perception reveals that farmers are deeply aware of the problems of land degradation in the region and have adopted various coping strategies that include changes in farming practices, reliance on multiple livelihoods and spatial reallocation of household labour (Pani and Mohapatra 2001; Pani and Carling 2013; Kumar and Pani 2013). Numerous works on ravine reclamation have been done by various boards under U.P., M.P., Gujarat and Rajasthan state governments and ministries of Government of India. The works done by Soil Conservation Department, Ministry of Agriculture, Forest Department and Planning Commission, Government of India, are to name a few.

1.1.9 Linkages Between Land Degradation and Rural Development

Rural development can only be done by the refining or improving the economic and social conditions of rural population. Further, India being an agrarian economy, land shortage and quality (in terms of productivity) as well as poverty is taken together that leads to low level of development (Reddy 2003). Then, rural land management becomes a very critical journey of innovation and development. Further that targets making science out of rural development, which will be extremely powerful system of a big community endeavour. Studies conducted in Italy taking consideration of the connection between changes in land vulnerability and

[3] SAC is Space Application Centre.

[4] ISRO (Indian Space Research Organisation)

socio-economic factors show the varied vulnerability level of land with certain socio-economic disparity in the region (Salvati and Zitti 2009; Salvati et al. 2011).

As per estimation, 1.5 billion is depended on degraded land for their livelihood in 2007, and 42% of world's very poor population live in degraded land (Nachtergale et al. 2010). In context to a developing country like India, rural development has valued as prime requisite in nation development as 68.84% of population is still rural (Census 2011). Well-being of human and land is multidimensional, and specifically poverty is regarded as both the reason and outcome of land degradation in rural areas (Gerber et al. 2014; Eswaran and Reich 1998). Further, Millennium Ecosystem Assessment has pointed out close relation in human development and ecosystem services taking consideration of four broad dimensions: security, health, basic living material and social relation (MEA 2005). Poverty is a subject of wide interpretation but focus at some specific attribute establishes its linkage with land degradation (Gerber et al. 2014).

In most of developing nations, rural poor are found in area prone to land and water degradation and have close association with poverty that links land as an asset (Barbier 2010; Gerber et al. 2014). As per one study, degradation can reduce yield as Africa faced mean 8.2% of fall in production due soil degradation, and there is an estimated 20% loss of productivity in Asia (Eswaran et al. 2001). As per one evaluation done by Aubreville (1949) in Africa, it is revealed that erosion and other processes are a consequence of mismanagement of land by poor farmer and were the prime factors for land degradation (Eswaran and Reich 1998). The space and time dimension is very important in this regard as land degrades slowly. Food security, environmental balance and land degradation are related, and rational understanding about soil, water and nutrient management and farmer perception is an important concern (Eswaran and Reich 1998).

Further, with 40% of the world's people living in the drylands, desertification is as much a human problem as an environmental problem and has a major impact on the quality of life of the entire society. It is important to get more precise statistics on the number of people who are directly affected by desertification and their capacity to adjust with its impacts. Assimilating desertification into mainstream development planning will ensure that improving the livelihoods of impacted people can be given the attention it deserves. Therefore, research in many dimensions suggests that dealing with desertification is not only implementing physical remedies, such as more 'sustainable land management', but requires social remedies too. This means that economic impacts and social impacts need to be dealt collectively (Low 2013).

The distribution of effects within a society should be assessed by examining impacts on different classes of people. Poverty, for example, can be both a cause and an effect of environmental degradation. Poverty-stricken people are prone to exhaust natural resources if they have no prospects of having access to other resources, and a degraded environment can increase disadvantage. Claims continue about a 'vicious circle' linking poverty to population growth, drought and land degradation (Cleaver and Schreiber 1993). Whether the poor are main instrumental for desertification or not, they certainly suffer from its effects, as their livelihoods entirely depend on the productivity of land (Hazell et al. 2002; Stringer 2009). Crucially in this context, the livelihoods of the majority of the rural poor depend on land (Nachtergaele et al. 2010). Land degradation in rural India has multifaceted impacts. One of the major aspects is its relation to socio-economic development (Pani and Carling 2013; Priya and Pani 2015, 2017; Pani 2016, 2017b). In rural areas, the poverty level varies state-wise. The increase in per capita income has a significant role in the agricultural and non-agricultural sectors. However, this is not only the factor, though is a key component to affect the livelihood in the rural areas (Nayyar 2006). Poverty coupled with high population densities have cited as key cause for land degradation (Bossio et al. 2010; Kangalawe and Lyimo 2010). The study by Fan et al. (2000) has inferred that improving poverty and development in the rural infrastructure enhances the productivity of the land. In turns, it might be lead to an increment in the purchasing power of the people in a country. A research by Wang et al. (2013) in China has suggested that the reduction in poverty is possible through the better agricultural practices.

1.1.10 The Context of the Study

Land degradation is considered to be one of the most severe global environmental challenges. It is an established fact that land degradation in any form always affects the productive capacity of land. Chambal ravine is perhaps the nucleus of one of the most severe types of land degradation that exists in India and of course is the least paid attention area. Land degradation has many facets in terms of scale, forms and associated environments since the word land itself encompasses soil, water, vegetation, landscape and climatic components of an ecosystem. Degradation is a process of change over time. Hence land degradation refers to a temporary or permanent decline in the productive capacity of the land, or its potential for environmental management. The lower Chambal valley of Madhya Pradesh is severely affected by gully erosion. The area is mostly dissected and is made up of steep ridges, low sloping hills, deep trenches and broad meanders. In the past few decades, the Chambal ravine has come to be seen as a national problem. The formation of ravines has been posing serious problems to the agricultural and irrigation planning of the area. The ravines are constantly damaging the valuable fertile land and are slowly engulfing the agricultural land every year and consequently have an impact on the socio-economic development of the region.

The most common land degradation problems in the area are:

- Soil erosion by water and soil loss
- Lack of sufficient vegetative cover
- Inefficient use of available water resources
- Agricultural land converted to ravine land
- High cost of production, low marginal return and out-migration of youth

Ravine encroachment of settlement and agricultural land are leading to shifting of settlement. For the last one decade, various methods have been employed by the government as well as local farmers' perspective to reclaim the ravine lands for the sustainable rural development in the Chambal Badlands area. However, these techniques have begun to turn the tides against gully erosion at places. Due to the torrential rainfall that is irregular and intensive, the gully erosion is inherently difficult to measure accurately due to problems of creating monitoring networks in large areas where rainfall and runoff are highly variable (Bull and Kirkby 2002: 11).

1.1.11 Conceptual Framework

This study addresses a range of questions, which cannot be answered through a narrow conceptual frame. Land degradation is a biophysical process that can be analysed by examining the natural processes, such as soil health and changes in it; but the causal interlinkages of land degradation remain incomplete without taking into account the social and economic processes that cause and are also influenced by land degradation. Thus, the conceptual framework that has been developed for the study combines both biophysical and socio-economic aspects of land degradation. The relationship between land degradation and socio-economic development is multidimensional and is often circular. On the one hand, land degradation adversely affects productivity of crop farming and hence restricts the possibility of an agriculture-led transformation of the rural economy, as suggested by John Mellor and others (Johnston and Mellor 1961). On the other hand, low levels of development, and lack of alternative livelihoods opportunities, force people to cultivate marginal and degraded land in an unsustainable manner and restrict their capacity to opt for sustainable land use practices. Land degradation, thus, is not only the cause of low levels of development; the causation might work in the other direction as well.[5] Further, land degradation primarily affects rural livelihoods through crop farming, but it is not the only channel through which the mechanism works (Fig. 1.1). As productivity of land declines, cost of production in crop farming declines, and farm profitability goes down; farmers also forced to look for alternative livelihood options. This process of involuntary diversification of livelihoods is often termed as 'distress diversification', and it typically involves a number of

[5]However, the land degradation-poverty-livelihood linkages are far from being uniform. Summarising the available evidence, Scherr has argued that while in some cases, poverty and natural resource degradation reinforce each other, causing a 'downward spiral'; in many other contexts, it has been found 'variously, that degradation resulted from natural forces rather than human mismanagement; indigenous technology developed to control degradation; local communities implemented land use controls to stabilise vegetative cover; or farmers diversified activities to reduce degradation while maintaining incomes (Forsyth et al., 1998). ... As the cost of land relative to labour increased, people often changed their methods of managing plants and animals and made land improvements to offset initial declines in productivity resulting from more intensive land use'(Scherr 2000:481).

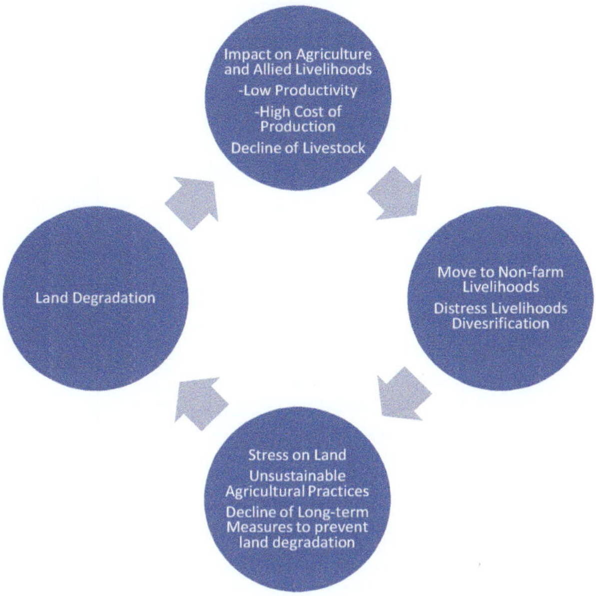

Fig. 1.1 Land degradation-livelihoods nexus. (Source: author's conceptualisation)

short-term, low-earning and environmentally damaging activities. The livestock and forest-based livelihoods also suffer because of the damages to natural resource base of the area, where land degradation continues to be present. Because of its uneven impacts on various livelihood assets and sectors of the rural economy, the socio-economic impacts of land degradation are uneven across different sections of the rural population.

In order to understand the implications of land degradation on sustainable rural development, the process of rural development has been analysed through its external and internal dynamics (Fig. 1.2). In the literature on rural development, the key external drivers of rural development include the urban stimulus and state interventions. The rural urban linkages can be through various kinds of flows in either direction. Capital, products or items of consumption generally flow in both directions, while labour flows out of rural to urban areas. These flows create demand for rural products in urban areas and act as a catalyst for investment and growth in the rural areas. Labour outflow results in inflow of remittances to the rural economy. State interventions similarly take various forms such as institutional changes (changes in the access to resources), capital investment, employment creation, infrastructure development and investment in human resource development. All these changes might bring changes in the levels of development in the rural areas.

At the foundations of development of rural areas is the natural resource base, such as land, water and forest. Land degradation affects the availability of natural resources in an area. Another significant dimension of the rural development is the population base and the demographic features of an area. Factors such as population

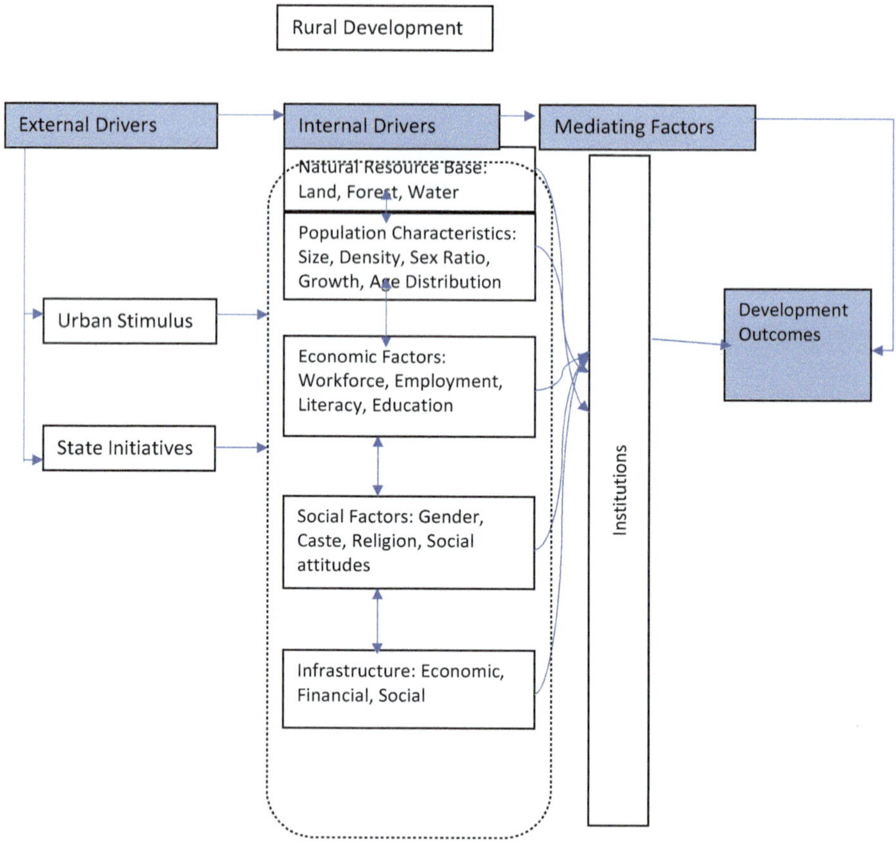

Fig. 1.2 Drivers of rural development. (Source: author's conceptualisation)

density, population growth, age and sex composition of the population reflect the levels of development as well as possible pathways for rural development. These factors, together with others, determine the economic structure of the area. The employment and livelihood profile of the area and human resource parameters such as education and health shape the process of development. Processes of development are also linked with various social institutions (such as caste, ethnicity and gender), which determine access (i.e. exclusion and inclusion) to various resources and opportunities. Finally, availability of infrastructure of various types, social, financial, communications, etc., is crucial for economic development. All these factors listed as internal drivers in the figure share two critical features. Firstly, all these factors are interrelated (e.g. natural resources, population and social factors determine the economic structure and are also shaped, to some extent, by it). Secondly,

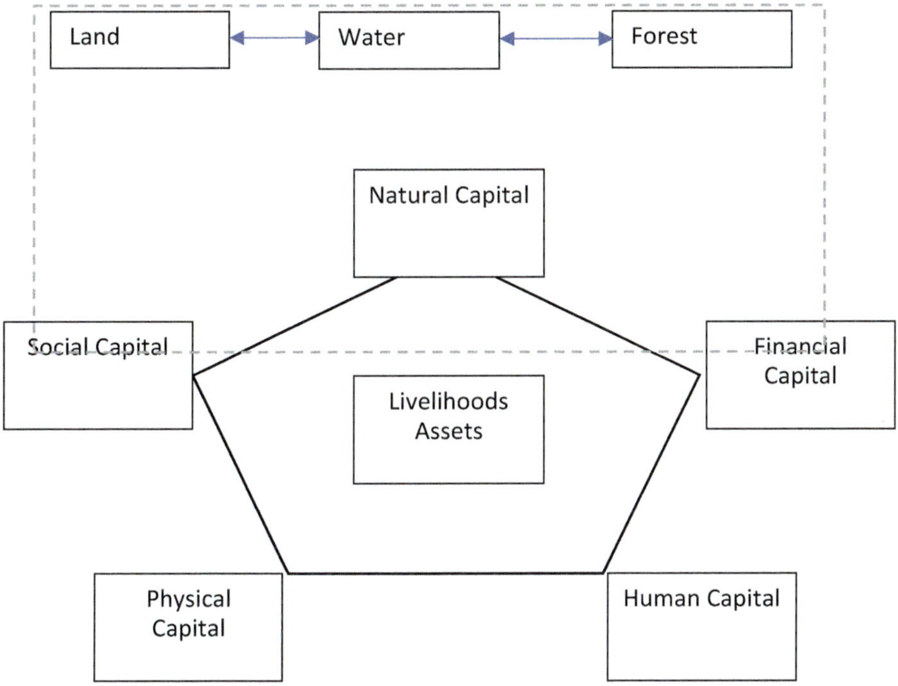

Fig. 1.3 Livelihood impacts of land degradation. (Source: author's conceptualisation based on literature survey)

all these factors act as catalysts for economic development; but these factors are also affected by the development process (Fig. 1.3).

The conceptual framework outlined here establishes a linkage with insights from the literature with the specific context of the study area. The lower Chambal valley is among the severely ravine-affected parts of India. Various types of human-driven landscape change have been noticed in the study area. It not only provides a scope for investigating the socio-economic factors responsible for the present-day land degradation scenario but also assesses the reclaimed ravine lands in the area. Geospatial investigations offer an excellent opportunity to study the land degradation status of a fragile environment. This is a relatively new technology that is currently being investigated by researchers and scientists with regard to the land degradation studies. Chambal valley of India provides a unique opportunity to understand the gully erosion in the form of land degradation. Hence it was felt to combine this technology to study the status of land degradation as well as the land reclamation, rapid change of land use/land cover practices and human interactions. The social forces that drive landscape constantly in the area along with the measures adopted for the sustainable rural development have been highlighted by the study. To sum up, this study attempts to examine the linkages of land degradation with socio-economic development of the ecologically fragile region, by using multiple data sets and combining geospatial and household survey-based research methods.

1.1.12 *Objectives, Database and Methodology*

1.1.12.1 The Key Questions

The primary objective of this book is to explore the interrelationship between land degradation and rural development in one of the regions severely affected by land degradation. The study attempts to examine the changes in the extent of ravines over time and space. The role of anthropogenic activities on ravines and the implications of land degradation for agriculture and on levels of rural development in the region are the key questions explored in this context. The coping mechanisms adopted by the farmers have been examined by studying the land reclamation, changes in agricultural practices and livelihoods. Based on these broad objectives, the broad questions that have been examined in this study include the following:

(i) What are the temporal and spatial dynamics of land degradation in the study area?
(ii) How is the land degradation interlinked with human-induced activities?
(iii) What are the implications of land degradation for agriculture and rural development?
(iv) What are the coping mechanisms that are being adopted by farmers and households in response to land degradation?
(v) What types of methods are adapted for the reclamation of the ravine lands and on which scale?
(vi) What are the programme and policies adopted to deal with the problems of land degradation in the study region, and how effective are they in overcoming the problems?

1.1.12.2 Database

To fulfil the objectives of the study, the following data sets have been used and generated:

1. Secondary Data

   a. i. Remote sensing data: Landsat-1, the Landsat Multispectral Scanner (MSS) of year 1974 with 60 m resolution, and the Landsat 8, the Operational Land Imager (OLI) 9 bands and the Thermal Infrared Sensor (TIRS) 2 bands (2014), with spatial resolution of 30 m, provided by US Geological Survey.
   ii. Periodic Google Images (1984–2014)
   b. The Survey of India Topographic Sheets (1984), scale 1:50,000

c. The district gazetteer, Population Census, Village Directory, Forest Survey of India, Human Development Reports, District Village Boundary Maps, NRSC wasteland data (1999, 2001, 2011), Agricultural Census, etc.
d. District- and village-wise population-related data will be provided by the Census of India, Office of the Registrar General and Census Commissioner, India (ORGI).
e. The Agricultural Census, Department of Agriculture and Corporation and Ministry of Agriculture, Government of India, New Delhi, will provide district and the tehsil level data, and other data related to agriculture is provided by Directorate of Economics and Statistics.

2. Primary Data

   i. Field survey: (a) physical field survey; (b) socio-economic household field survey and (c) focus group discussions analysis. The details of the field survey are presented below.

1.1.12.3 Methodology

Land degradation is a complex biophysical process. To understand the land degradation processes as a whole, it needs to be evaluated through physical and socio-economic data analysis. As it is mentioned earlier, this project has been based on combination of geospatial technology along with physical and socio-economic field survey and other secondary data analyses.

To fulfil the objectives of the study, the following methodology have been adopted.

Firstly, to understand the temporal and spatial dynamics of land degradation, the remote sensing digital satellite images of 1974 and 2014 and Survey of India topographic sheets of 1984 have been used along with google image from 1984 periodically. To capture the spatial dynamics of land degradation for the entire district as the degraded land covered a large area, the aerial extent of degraded land has been mapped using ERDAS and ARC GIS software. The expansion, the nature of expansion and the direction of gully extension as well as extinction have been delineated using online-based visual image interpretation techniques using tone, texture, shape, size, pattern and other keys to identify ravine and other classes in ERDAS and ArcGIS software.

The nature and degree of erosion also mapped using visual image interpretation techniques, the mixing of spectral bands created confusion during classification of barren land and non-agricultural land which are rectified with the repetitive field surveys, and all the analyses were crossed-check with the ground truth verifications. The difficulties to distinguish the fellow land and shallow ravine land in the remote sensing data in some areas were minimised using field observation and ground truth verifications. The visual interpretation keys have been used to identify and separate the mixture land impression in many places. In the study area, the nature of gullies is very rugged; therefore the texture and tone have taken consideration for visual-based interpretation to classify the gullies and differentiated the gully and non-gully areas.

To capture the temporal change of four decades of gully erosion (1974–2014), the 2014 degraded map have been superimposed on prepared land degraded map of year 1974 using ARC GIS version 10.3 software and have been successfully delineated and calculated the total change of degraded areas over 40 years.

The observed changed areas were compared with the Google Earth images, and the changes were digitised manually as these are very small in size in many places. The reclaimed land or levelled lands were digitised precisely because of the nature of the land, and it was cross-checked with the high-resolution google images and by repetitive field surveys.

In order to assess the present-day land degradation process and development across space, various thematic layers such as land use/land cover maps, ravine/gully change maps, the types of ravine maps, reclaimed ravine land (which are presented here as levelled land), drainage density and Normalised Difference Vegetation Index (NDVI) maps, agricultural maps, etc. have been generated based on various remote sensing image analysis, along with Survey of India topographic sheets and google images. The total areas of different land use classes and other layers were also calculated, and inter-temporal changes were computed. Drainage density is the length of primary streams per unit area and is a commonly used index of erosion intensity in generalised erosion hazard assessment (Morgan 2009). A drainage density map of the area has been prepared from ASTER Digital Elevation Model Maps using ARC GIS software to mark the low medium and high-density areas. Normalised Difference Vegetation Index (NDVI) is a measure of the contrast in reflectance of near-infrared and red bands. The NDVI maps have been prepared from the LANDSAT 82014 images. NDVI data cannot automatically be interpreted for identifying the land degradation, but can be considered for operational use to link with the vegetation cover and land degradation. In order to understand the erosional processes, the impact of erosion and its on-site and off-site implications were captured in field photographs of different seasons. The photographs have been used to understand the ravine formations and implications evidently.

The study has concentrated on the identification, estimation and spatial distribution of ravines/gully degradation and the reclaimed ravine lands in the Chambal valley. In order to have the idea of the processes acting on the surface of the earth and their scale, different resolution images will be used. Detailed field work has been conducted with the global positioning system, laser distance metre, along with other equipment in order to gather the information on the human interaction and activities in the ravenous zones. These maps were used for further analysis to understand the ravine status of the study district. Finally, the ground truth verifications and field observations have been done in different periods in 2014, 2015, and 2016, respectively, and final maps have been prepared.

Identification of Gully-Affected and Non-affected Villages

In order to understand the status of villages in terms of gully affected villages and non-gully affected villages, the village boundary of the study districts (a total number of villages are 794 as per census) has been digitised using ARC GIS software from the District Village Boundary Maps, and all the villages were superimposed on the prepared ravine maps of 2014. Villages which are situated inside the gully-/ravine-affected area were classified as ravine-infested villages, and the rest of villages, which are outside of the ravine area, were classified as non-affected villages. In this process, status of each village has been identified. The nature of each village, their location, status and other socio-economic information were added with the individual village and also were evaluated using village census data of 2011 in GIS environment.

Socio-economic data has been collected from the villages and analysed to assess the extent and implications of land degradation for livelihoods in the area. In order to understand the interrelationships between land degradation and socio-economic development across space, this study has tried to combine two distinct types of data sources in an innovative way. Based on this mapping exercise outlined above using remote sensing and GIS, the villages of the district were classified into two groups: those located within the degraded area and those situated outside (as explained above). Finally, seven villages were selected on the basis of extent of degradation, one each from highly degraded, moderately degraded and less degraded zones identified through remote sensing data. Since villages generally are relatively small in the region, a sample of households of the selected villages was surveyed. The household level survey has cover employment, earnings, agricultural production, cropping pattern, input-use and labour-use, etc. The focus was to explore the implications of land degradation for agriculture at the household level. The coping mechanism adopted by the local people and programmes implemented by government and non-governmental organisations were also studied selectively. Finally, the evidences and information obtained for the socio-economic factors and the interactions of anthropogenic activities have been assessed through GIS. The GIS analysis has been used bringing out not only the present-day scenario of land degradation but also the reclaimed ravine lands in the study area.

Sampling Design

The study area has a lot of peculiarities in terms of the ecological and socio-economic characteristics. Given the objectives of the study, a multistage sampling design was followed. In the first stage, all the 198 villages in the region were categorised into three categories as per their spatial location – those located in the high, moderate and low degraded zones. This categorisation was based on the basis of remote sensing data analysis. Given the distribution of the villages in these three categories, seven villages were purposively chosen from among the degraded villages. In selecting the villages, apart from the extent of land degradation, two additional characteristics were also taken into account: (a) the villages with a history of land reclamation and (b) villages near the river Chambal. The justification for these two criteria is as follows. Since one of the objectives of the study is to examine coping strategies with respect to land degradation, it was important to include villages with a history of land degradation. Information regarding history of land degradation in the area was collected through interview of key informants. Secondly, since the land degradation process is known to be more active in areas nearer to the Chambal River, villages nearer to the river were purposively chosen.

After the choice of the villages, the following steps were followed for the selection of the sample households. Since the objective of the primary survey was primarily to examine the interlinkages between land degradation and livelihoods in agriculture and allied activities, the first step was to identify a zone of degraded or levelled agricultural land in the ravine area. This was done through the physical verification of the agricultural land in the areas within a vicinity of five kilometres of the village. Since there are multiple patches of degraded land which are under cultivation, typically one, relatively large, patch of agricultural land was selected randomly from the different patches of degraded agricultural land in the villages. Through key informants in the villages, the cultivator households cultivating in the degraded land, irrespective of the legal status of ownership of land, were identified and selected for the detailed household level survey. In the cases that the cultivating households could not be contacted or were not interested to participate in the survey, a substitute was included from the nearest agricultural land cultivated. An attempt was made to include all the cultivating households cultivating in a particular zone of degraded land. The sampling framework was designed keeping in mind the objective of the exercise, i.e. to investigate the interrelationship between land degradation and livelihoods through agricultural activities. The stepwise sampling method ensured that the sample households represent the sample of those households who have been affected by land degradation to different degrees. A limitation of the sampling exercise was that it might not include various non-agricultural activities and those who were not involved in agricultural operations. In order to overcome these limitations, focus group discussions were conducted in each of the selected villages with three separate groups: (a) land less agricultural labour households; (b) women in the working-age group (15–59); and (c) those who are partially or fully dependent upon non-agricultural livelihoods.

1.1.13 Organisation of the Book

The first chapter is an introduction of the theme and a global view which dealt with all aspects of land degradation, related to environmental degradation and development, links between land degradation and development; meaning, definition and the types of land degradation; causes, processes and measurements of land degradation; the land degradation scenario in India; and the linkage between land degradation and rural development. The objectives, database and adopted methodology for the study are also placed in this section. The second chapter is on the spatial and temporal dimensions of land degradation in the degraded Chambal valley. In this chapter, the land degradation status, its extent over the years and the present condition have been discussed. This chapter, primarily relying upon remote sensing and GIS techniques, presents a detailed examination of land degradation in the study area, along with its linkages with the existing geology, geomorphology, soil, forest resources climate, etc. The third chapter is focused on the socio-economic profile of the study region, and the fourth one covers the linkages between land degradation, rural development and livelihoods. Largely based on secondary data sources, a comparative analysis of the socio-economic scenario of the degraded and non-degraded villages in the district has been placed in the fourth chapter. The social, economic and educational aspects of development are mapped, and an attempt has been made to establish a link between the land degradation and rural development. The fifth chapter is based on primary survey to analyse the rural development and land degradation. In this chapter, the data of physical survey and the household analysis have been used to get the ground reality of the ravine-affected villages. The conclusion and summary of the findings have been placed in the last chapter.

References

Adams CR, Eswaran H (2000) Global land resources in the context of food and environmental security. In: Advances in land resources management for the 20th century. Soil Conservation Society of India, New Delhi

Bai ZG, Dent DL, Olsson L, Schaepman ME (2008) Proxy global assessment of land degradation. Soil Use Manag 24(3):223–234. Wiley Online Library

Barbier EB (2010) Poverty, development, and environment. Environ Dev Econ 15(6):635–660. Cambridge University Press

Barrow CJ (1991) Land degradation: development and breakdown of terrestrial environments. Cambridge University Press, Cambridge

Besley T (1995) Property rights and investment incentives: theory and evidence from Ghana. J Polit Econ 103(5):903–937. The University of Chicago Press

Blaikie P (1985) The political economy of soil erosion in developing countries. London and New York: Longman

Blaikie, P, Brookfield H (2015) Land degradation and society (Blaikie P, Brookfield H, eds). Routledge, New York

Bonilla CA, Johnson OI (2012) Soil erodibility mapping and its correlation with soil properties in Central Chile. Geoderma 189:116–123. Elsevier

Bossio D, Geheb K, Critchley W (2010) Managing water by managing land: addressing land degradation to improve water productivity and rural livelihoods. Agric Water Manag 97(4):536–542

Brown LR, Kane H (1995) Full house: reassessing the earth's population carrying capacity. Earthscan, London

Bull LJ, Kirkby MJ (2002) Channel heads and channel extension. In: Bull LJ, Kirkby MJ (eds) Dryland river: hydrology and geomorphology of semi-arid channels. Wiley, Chichester, pp 263–287

Census of India (2011) Census of India 2011, series – 24 Part Xii-A, village and town directory, 2011, directorate of census operations, Madhya Pradesh, Morena (Part A and B). Available at http://censusindia.gov.in/2011census/dchb/DCHB_A/23/2302_PART_A_DCHB_MORENA.pdfandhttp://censusindia.gov.in/2011census/dchb/2302_PART_B_DCHB_MORENA.pdf

Chatterjee RS, Saha SK, Kumar S, Mathew S, Lakhera RC, Dadhwal VK (2009) Interferometric SAR for characterization of ravines as a function of their density, depth, and surface cover. ISPRS J Photogramm Remote Sens 64(5):472–481

Cleaver K, Schreiber G (1993) The population, agriculture and environment Nexus in Sub-Saharan Africa, Agriculture and rural development series. World Bank, Washington, DC

Contador JF, Lavado S, Schnabel A, Gutiérrez G, Pulido Fernández M (2009) Mapping sensitivity to land degradation in Extremadura. SW Spain. Land Degrad Dev 20(2). Wiley Online Library:129–144

Crosson P (1994) Degradation of resources as a threat to sustainable agriculture. In first world congress of professionals in Agronomy, Santiago, Chile, pp 5–8

Dobie P (2001) Poverty and the drylands. UNDP report. Available at: http://www.undp.org/content/dam/undp/library/Environment%20and%20Energy/sustainable%20land%20management/The%20Global%20Drylands%20Initiative,%202001-09%20-%20Challenge%20Paper-%20Poverty%20and%20the%20Drylands.pdf

Dregne HE, Chou NT (1992) Global desertification dimensions and costs. Degradation and restoration of arid lands. International Center for Arid and Semiarid Land Studies, Texas Tech University, Lubbock, TX, USA

Eswaran H, Reich P (1998) Desertification: a global assessment and risks to sustainability. In: Proceedings of the 16th international congress of soil science, Montpellier, France

Eswaran H, Lal R, Reich PF (2001) Land degradation: an overview. In: Bridges EM (ed) Responses to land degradation. Oxford & IBH Publishing, New Delhi, pp 20–35

Fan S, Hazell P, Haque T (2000) Targeting public investments by agro-ecological zone to achieve growth and poverty alleviation goals in rural India. Food Policy 25(4):411–428

FAO (1996) The state of food and agriculture, 1996. FAO, Rome

FAO (2005) The state of food and agriculture 2005. FAO, Rome

Feng J, Wang T, Qi S, Xie C (2005) Land degradation in the source region of the Yellow River, Northeast Qinghai-Xizang Plateau: classification and evaluation. Environ Geol 47(4):459–466. Springer

Gao J, Liu Y (2010) Determination of land degradation causes in Tongyu County, Northeast China via land cover change detection. Int J Appl Earth Obs Geoinf 12(1):9–16. Elsevier

Geist HJ, Lambin EF (2004) Dynamic causal patterns of desertification. AIBS Bull 54(9). American Institute of Biological Sciences:817–829

Gerber N, Nkonya E, von Braun J (2014) Land degradation, poverty and marginality. In: Von Braun J, Gatzweiler FW (eds) Marginality: addressing the Nexus of poverty, exclusion and ecology. Springer, Berlin, pp 181–202

Grepperud S (1997) Soil conservation as an investment in Land. J Dev Econ 54(2):455–467. North-Holland

Hazell P, Jansen H, Ruben R, Kuyvenhoven A (2002) Investing in poor people in poor lands. IFAD/IFPRI/NIFP/Wageningen University and Research Centre/International Food Policy Research Institute, Washington, DC. http://www.fao.org/docrep/003/w1358e/w1358e.pdf

ISRIC & FAO (2010) Harmonized world soil database (version 1.1). Food and Agriculture Organization, Rome, Italy and International Institute for Applied Systems Analysis, Laxenburg, Austria

Johnston BF, Mellor JW (1961) The role of agriculture in economic development. Am Econ Rev 51(4):566–593

Kabubo-Mariara J (2007) Land conservation and tenure security in Kenya: Boserup's hypothesis revisited. Ecol Econ 64(1):25–35. Elsevier

Kangalawe RY, Lyimo JG (2010) Population dynamics, rural livelihoods and environmental degradation: some experiences from Tanzania. Environ Dev Sustain 12(6):985–997

Kumar H, Pani P (2013) Effects of soil erosion on agricultural productivity in semi-arid regions: the case of lower Chambal Valley. J Rural Dev 32(2):165–184

Lal R (2001) Soil degradation by erosion. Land Degrad Dev 12(6):519–539. https://doi.org/10.1002/ldr.472

Low PS (2013) Economic and social impacts of desertification, land degradation and drought. In White paper I, UNCCD 2nd scientific conference

Marzolff I, Pani P (2017) Dynamics and patterns of land levelling for agricultural reclamation of erosional badlands in Chambal Valley (Madhya Pradesh, India). Earth Surf Process Landf. https://doi.org/10.1002/esp.4266

MEA (Millennium Ecosystem Assessment) (2005) Ecosystems and human well-being. Island Press, Washington, DC

Ministry of Agriculture (1984) Report of the working group on reclamation and development of ravines for formulation of the five year plan, New Delhi

Morgan RPC (2009) Soil erosion and conservation. Wiley, New York

Mythili G, Goedecke J (2016) Economics of land degradation in India. In: Economics of land degradation and improvement – a global assessment for sustainable development. Springer, Cham, pp 431–469

Nachtergaele F, Petri M, Biancalani R, Van Lynden G, Van Velthuizen H, Bloise M (2010) Global Land Degradation Information System (GLADIS). Beta version. An information database for land degradation assessment at global level. Land degradation assessment in drylands technical report, p 17

Nayyar D (2006) Economic growth in independent India. Econ Polit Wkly 41(15):1451–1458

Nkonya E, Pender J, Kaizzi KC, Kato E, Mugarura S, Ssali H, Muwonge J (2008) Linkages between land management, land degradation, and poverty in Sub-Saharan Africa: the case of Uganda, vol 159. Intl Food Policy Res Inst, Washington, DC

Nkonya E, Winslow M, Reed MS, Mortimore M, Mirzabaev A (2011) Monitoring and assessing the influence of social, economic and policy factors on sustainable land management in Drylands. Land Degrad Dev 22(2):240–247. https://doi.org/10.1002/ldr.1048

NRSA (1995) Wasteland atlas of India. National Remote Sensing Agency, Hyderabad

NRSA (2011) http://bhuvan.nrsc.gov.in/gis/thematic/tools/document/wl/WL0809.pdf

Oldeman LR, Hakkeling RTAU, Sombroek WG (1991) World map of the status of human-induced soil degradation: an explanatory note. Global Assessment of Soil Degradation (GLASOD), Wageningen. ISRIC/UNEP/WSC/ISSS/FAO/ITC

Pani P (2016) Controlling gully erosion: an analysis of land reclamation processes in Chambal Valley, India. Dev Pract 26(8):1047–1059

Pani P (2017a) Ravine erosion and livelihoods in semi-arid India: implications for socioeconomic development. J Asian Afr Stud 53(3):437–454

Pani P (2017b) Chambal without ravines. Down Earth 26(8):80–81

Pani P, Carling P (2013) Land degradation and spatial vulnerabilities: a study of inter-village differences in Chambal Valley, India. Asian Geogr 30(1):65–79. https://doi.org/10.1080/10225706.2012.754775. Routledge

Pani P, Mohapatra SN (2001) Delineation and monitoring of gullied and ravinous lands in a part of lower Chambal Valley, India, using remote sensing and GIS. 22nd Asian conference on remote sensing, vol 54, pp 5–9

Pani P, Mohapatra SN (2011) Mapping and assessment of land degradation using NDVI techniques in a part of Chambal Valley. Indian J Landsc Syst Ecol Stud 34(2):155–160

Pani P, Mohapatra SN, Singh UC (2005) Evaluation of different landform units and Neotectonic implications in a part of lower Chambal Valley, using satellite images – a computer aided geomorphic approach. In: Bhattacharya AR, Agarwal KK (eds) Himalayan Orogen-foreland interaction. Palaeontological Society of India, Lucknow, pp 219–224

Pani P, Mohaptra SN, Ranga V (2008) Assessment of gully erosion in a part of lower Chambal Valley using digital image processing and GIS environment. Bull Geogr Sci 3(1):23–26

Pender JL, Kerr JM (1998) Determinants of farmers' indigenous soil and water conservation investments in semi-arid India. Agric Econ 19(1–2):113–125. Elsevier

Pimentel D, Harvey C, Resosudarmo P, Sinclair K, Kurz D, McNair M, Crist S, Shpritz L, Fitton L, Saffouri R, Blair R (1995) Environmental and economic costs of soil erosion and conservation benefits. Science 267(5201):1117–1123. American Association for the Advancement of Science

Pokharel T (2015) Human sustainable development: global issues. In: 13th South Asia management form (Colombo, Sri Lanka), pp 26–27. Colombo

Priya R, Pani P (2015) Land degradation and deforestation in India: a district level analysis. ANNALS of the National Association of Geographers, India 35(1):50–67

Priya R, Pani P (2017) Land degradation and agricultural productivity: a district level analysis, India. J Rural Dev 36(4):557–568

Reddy VR (2003) Land degradation in India: extent, costs and determinants. Econ Pol Wkly 38(44):4700–4713

Ross A (2009) Modern interpretations of sustainable development. J Law Soc 36(1):32–54

SAC, ISRO (2007) Desertification & land degradation Atlas of India. Space Application Centre/ISRP, Ahmedabad

Safriel UN, Adeel Z (2005) Dryland systems. In: Hassan R, Scholes R, Ash N (eds) Ecosystems and human well-being: current state and trends, vol 1. Island Press, Washington, DC, pp 623–662

Salvati L, Zitti M (2009) Assessing the impact of ecological and economic factors on land degradation vulnerability through multiway analysis. Ecol Indic 9(2):357–363

Salvati L, Mancini A, Bajocco S, Gemmiti R, Carlucci M (2011) Socioeconomic development and vulnerability to land degradation in Italy. Reg Environ Chang 11(4):767–777. Springer

Salvati L, Mavrakis A, Colantoni A, Mancino G, Ferrara A (2015) Complex adaptive systems, soil degradation and land sensitivity to desertification: a multivariate assessment of Italian agro-forest landscape. Sci Total Environ 521:235–245. Elsevier

Scherr SJ (2000) A downward spiral? Research evidence on the relationship between poverty and natural resource degradation. Food Policy 25(4):479–498. Elsevier

Scherr, SJ, Yadav S (1996) Land degradation in the developing world: implications for food, agriculture, and the environment to 2020 (No. 14). Washington, DC

Schröder JJ, Schulte RPO, Creamer RE, Delgado A, Van Leeuwen J, Lehtinen T, Rutgers M, Spiegel H, Staes J, Tóth G (2016) The elusive role of soil quality in nutrient cycling: a review. Soil Use Manag 32(4):476–486. Wiley Online Library

Sharma HS (1980) Ravine erosion in India. Concept Publishing Company, New Delhi

Snel M, Bot A (2003) Draft paper proposed indicators for land degradation assessment of drylands

Soubbotina TP (2004) Beyond economic growth: an introduction to sustainable development. The World Bank, Washington, DC

Stephens PR, Chilar J (1981) The potential of remote sensing to monitor soil erosion on cropland. Proc Fifteenth Int Symp Rem Sem Enriron, ERIM, Ann Arbor.

Stephens PR, Chilar J (1982) Remote sensing based methodology to map and monitor cropland erosion. Proc Seventh Canadian Symp Rem Sens

Sternberg T (2008) Environmental challenges in Mongolia's dryland pastoral landscape. J Arid Environ 72(7):1294–1304. Elsevier

Stringer LC (2009) Testing the orthodoxies of land degradation policy in Swaziland. Land Use Policy 26(2):157–168. Elsevier

TERI (1998) Looking Back to think ahead: GREEN (Growth with resource enhancement of environment and nature) India- 2047. TERI, New Delhi

UNCED (1992) United Nations Conference on Environment and Development (UNCED), Rio de Janeiro, 3–14 June 1992. https://sustainabledevelopment.un.org/content/documents/Agenda21.pdf

UNEP (1992) United Nations Environment Programme (1992) world atlas of desertification. London, p 69

UNEP (2008) Africa: atlas of our changing environment. Nairobi, Kenya: United Nations Environment Programme, Division of Early Warning and Assessment (DEWA). EarthPrint, 390 p. On line at http://www.unep.org/dewa/africa/AfricaAtlas/

UNSO (United Nations Sudano-Sahelian Office) (ed) (1993) 19P9a4st.oral develop- ment in Africa. Proceedings of the first technical consultation of donor and international development agencies, December 1993, Paris. New York

von Braun J, Gerber N, Mirzabaev A, Nkonya E (2013) The economics of land degradation. ZEF working papers 109. Bonn

Voortman RL, Sonneveld BGJS, Keyzer MA (2003) African land ecology: opportunities and constraints for agricultural development. Ambio JSTOR:367–373

Wadia FK (1996) Agro-climatic regional planning at zone level. In: Basu DN, Guha GS (eds) Agro-climatic regional planning in India, Part one: concepts and applications. Concept Publishing Company, New Delhi, pp 85–134

Wang F, Pan X, Wang D, Shen C, Lu Q (2013) Combating desertification in China: past, present and future. Land Use Policy 31:311–313. Elsevier

Wischmeier WH (1976) Use and misuse of the universal soil loss equation. J Soil Water Conserv 31(1):5–9

World Bank (2010) Development and climate change–world development report 2010. The World Bank, Washington, DC

Chapter 2
Land Degradation in Chambal Valley: Spatial and Temporal Dimensions

Abstract The second chapter is on the spatial and temporal dimensions of land degradation in the degraded Chambal valley. In this chapter, the land degradation status, its extent over the years and the present condition have been discussed. This chapter, primarily relying upon remote sensing and GIS techniques, presents a detailed examination of land degradation in the study area, along with its linkages with the existing geology, geomorphology, soil, forest resources climate, etc.

Keywords Causes of land degradation · Land degradation processes · Types of land degradation · Types of ravine · Chambal ravine · Land use/land cover change

2.1 Introduction

Land degradation is one of the key environmental issues in India. The gully and ravine erosions are the most significant category of land degradation problem in India. Nearly 0.32% of the total geographical area (TGA) of the country is affected by gullies and ravines in Indian subcontinent (NRSC 2008–2009).[1] It was always a concern in 1943 where the National Planning Committee estimated that around 3.8% of United India's land area were affected by gully severely and that 8% of the India's United Provinces (approximately 2 million ha) were ravines which were without the vegetation cover. After independence, as per the National Commission On Agriculture (1976), around 3.67 mha[2] land was affected by the ravine. As per the government of India (1972), five major ravine erosion zone had been identified on

[1] National Remote Sensing Centre, India Space Research Organisation.

[2] Mha is million hectors.

P. Pani, *Land Degradation and Socio-Economic Development*, Advances in Asian Human-Environmental Research,
https://doi.org/10.1007/978-3-030-42074-1_2

the basis of their severity in India (Figure No 1). Sharma (1980) has also identified the same. The identified zones are as follows:

(i) Zone of very severe ravine erosion:

The Yamuna-Chambal ravine zone: The largest Ravine Zone is the Yamuna-Chambal. In this zone nearly 5000 Km^2 area have been affected by the Ravines.

(ii) Zone of severe ravine erosion:

In arid Gujarat, the Ravine extends from the southern bank of the Tapti River in Rajasthan border. Apart from Tapti, Watrak, Sabarmati, Mahi, Narmada and their tributaries are affected by the severe ravine erosion. The part of western sub-Himalayas also comes under this zone along with Siwalik foot hills belt. The Chota Nagpur Plateau ravine nature is also severe type ravine-affected zone.

(iii) Zone of moderate ravine erosion:

The tract of Ganga plain, the valleys of Ramganga and Sai and some part of Godavari valley in Maharashtra and Andhra Pradesh and Kathiawar experiencing moderate ravine erosion.

(iv) Zone of insignificant ravine erosion:

The whole of Decan, south of Godavari, east of Varanasi, western Rajasthan, have been facing the problem but in a small scale.

The entire description depicts that the Indian lands are prone to gully and ravine erosion but in a different ecological zone.

In this chapter the attempt has been made to analyse the land degradation in the form of gully and ravine erosion, its expansion over four decades and its causes and classifications. The study has been carried out in one of severely affected ravine district Morena a part of Chambal region.

2.2 Material and Methods

The following sources and materials have been used in this study:

- The base map prepared from topographical map of Survey of India 1984.
- Village boundary map prepared from District Census Village Boundary Map using ArcGIS software. (Details of description of methodology have been described in Chap. 1).
- Land use/land cover maps of 1974 and 2014 and ravine maps of 1974 and 2014 have been prepared using Landsat 1 (Multispectral Scanner (MSS) sensor) image and Landsat 8 (Operational Land Imager (OLI) sensor) images for the years 1974 and 2014, respectively, using ERDAS and ArcGIS software. (For details see Chap. 1.)
- Ground truth verifications and field observations have been done in 2014, 2015 and 2016, respectively, using various field instruments, described in Chap. 1.

- Field photographs of different seasons have been used to understand the ravine formations and implications.

The geo-database has been prepared in ArcGIS environment comprising of various thematic maps such as land use, land cover, ravines extension map, NDVI maps, drainage density map, ravine classification maps, district boundary, village boundary, etc. These thematic maps have been prepared using onscreen visual interpretation techniques for different year images of 1974 and 2014 using ArcGIS software. The change of total degraded areas, the different land use classes, etc. also calculated and compared the values.

2.3 Introduction to the Study Area

The study area is one of the severely ravine-affected Morena districts in Chambal Region. The study area is located between the geographical coordinates of $26^0 17'15''$N to $26^0 52'22''$N latitude and $76^0 28'30''$E to $78^0 32'55''$E longitude. The geographical area covers near about 4989 km^2. The region is well drained by the river Chambal, one of the major tributaries of Yamuna river (Fig. 2.1). The drainage of the area is mainly controlled by river Chambal flowing in SW-NE direction and its tributary Kunwari, which flows from south to north and northeast and meets the Chambal River on its right bank. Most part of the Chambal valley is highly dissected and inaccessible. Steep ridges, low-sloping hills, deep trenches and broad meanders are the common features of the study area (Pani 2017a). Soils of the area are broadly of two types: one category is sandy loam to loam with low phosphorous and salt content and the other category is clayey loam with low phosphorous and salt content (GoMP 1996).

In terms of land degradation, Chambal region is among the worst affected region of India due to its nature of severity and its vastness. Ravine erosion is a major concern for the region developing a major physical and economic implication (Carling and Pani 2013). In this region gullies are widely spread and form extensive network of ravine. Ravine refers to deep gullies running parallel to subparallel to each other. It is a narrow steep-sided V-shaped valley and larger than a gully. The implications of gully erosion are not only limited to the loss of agricultural land; in numerous localities, it has also caused for shifting of villages either in a village level or in household level (Pani and Mohapatra 2001). The ravine development and expansion are having significant impact on environment of the surrounding region, which leads to large volume of soil loss after every rainy season. It has been observed that topsoil erosion has resulted in loss of productive capacity of agricultural land in long run. This semiarid region Chambal is a comparatively less developed part of India. The area is still dependent on primitive occupation like agriculture, around 80% of the rural people of this region are dependent on it. Several unsuccessful efforts to control gully and ravine erosion and many ways of land reclamation have been introduced since the early 1960s by government agencies, farmers and locals. The part of three large states UP, Rajasthan and M.P. are affected by the land degradation problem by

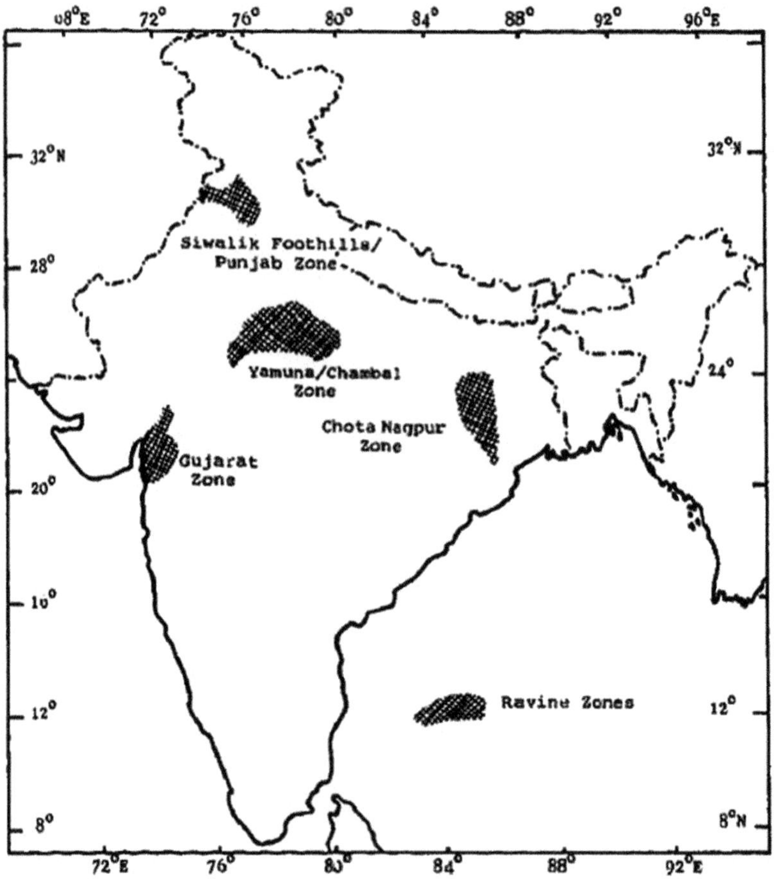

Fig. 2.1 Four major ravine zones in India. (Source: Adapted from GoI (1972) cited in Sharma (1980))

severe ravine erosion problem. Morena district is one of the ravine-affected districts in M.P. situated in the valley. It is important to note that neither is Morena, the district headquarters far from the state capital, nor is it a remote part- the National Highway no. 3 that connects Bhopal the state capital with Delhi and railway lines run through it. But, because of its typical geographical features, it is considered to be a rather 'backward' area.[3] This makes it a suitable area to study the implications of land degradation on socio-economic development. However, due to the difficult terrain and inaccessibility of the region, very limited field-based research has been

[3]This area known as Chambal Badlands is well-known for its long history of high crime rates and its remoteness. In the 1970s and early 1980s, the area was known to the world for the hideouts of bandits. The nature of inaccessibility of the ravines was used as a shelter for criminals (Pani 2017). The details of the socio-economic conditions of the study area has been discussed in the following chapter.

done in this area so far, particularly to understand the characteristics of ravine and its expansion, which need periodic observation and monitoring.

The ravine land is generally devoid of a good vegetation cover. Most of the area is covered with thorny bushes and sparse vegetation. As per the State of Forest Report 2011, the district has a forest cover of 14.63% as against the state average of 25.21%. Of the total area under forest cover in the district, there is no area under dense forest, moderately dense forest account for 13.42%, and open forest account for 90.29% of the total forest area. A further 8.22% of the total geographical area is under scrub (Forest Survey of India 2011). As already stated, land degradation by ravines and gullies is widespread in India, but the most critical zone is the Chambal valley (Pani 2017a). The area is densely populated, and a large number of this population live in the rural areas and are dependent on agriculture. The region has a long history of drought, occasional excessive rains and food shortages (GoMP 1996:118–119). The high rate of soil loss due to topsoil erosion, encroachment of gullies and ravine formation is a severe threat for the overall development for the region (Pani 2016a). Of the gross cropped area, food crops account for 55.65%, and non-food crops account for 44.35% of the area in 2007–2008.[4] During 2005–2008, nearly 66% of the net sown area has been under irrigation. Canals, tube wells and wells are the main sources of irrigation in the area. Ravines are one of the causes for damaging and engulfing fertile land in India. Gully and ravines are considered to be the least productive area in study region. It has been estimated that ravines are destroying productive tablelands approximately at the rate of 8000 ha per year in India, losing thereby the food production to the tune of 3 million tonnes apart from fuel wood and fodder (Das 1985a, b; Kandrika and Dwivedi 2013). An estimated 3.98 million ha of ravines appear along the rivers Yamuna, Chambal, Mahi, Sabarmati and others (Ministry of Agriculture 1984) (Fig. 2.2).

However, rainfed agriculture in the region is characterised by low productivity and low levels of marketisation. The region has a long history of drought, occasional excessive rains and food shortages (GoMP 1996: 118–119). As against the rural headcount poverty ratio of 36.8% in Madhya Pradesh, the poverty ratio in Morena was 20.8% in 2004–2005, while the urban poverty ratio was 42.1 in Morena as against the state level urban poverty ratio of 42.7 (Chaudhury and Gupta 2009; GoMP 2007).

2.3.1 Physiography

River Chambal, one of the major tributaries of river Yamuna, originated from the Vindhyan range near Mhow in Indore District of Madhya Pradesh. Physiographically the area to the north of river Chambal is characterised by deeply dissected plateau and developing undulating topography. It has been believed that the Chambal River is

---

[4] Since 1974–1975 there has been a phenomenal growth of area under winter crops (*rabi*) in the district.

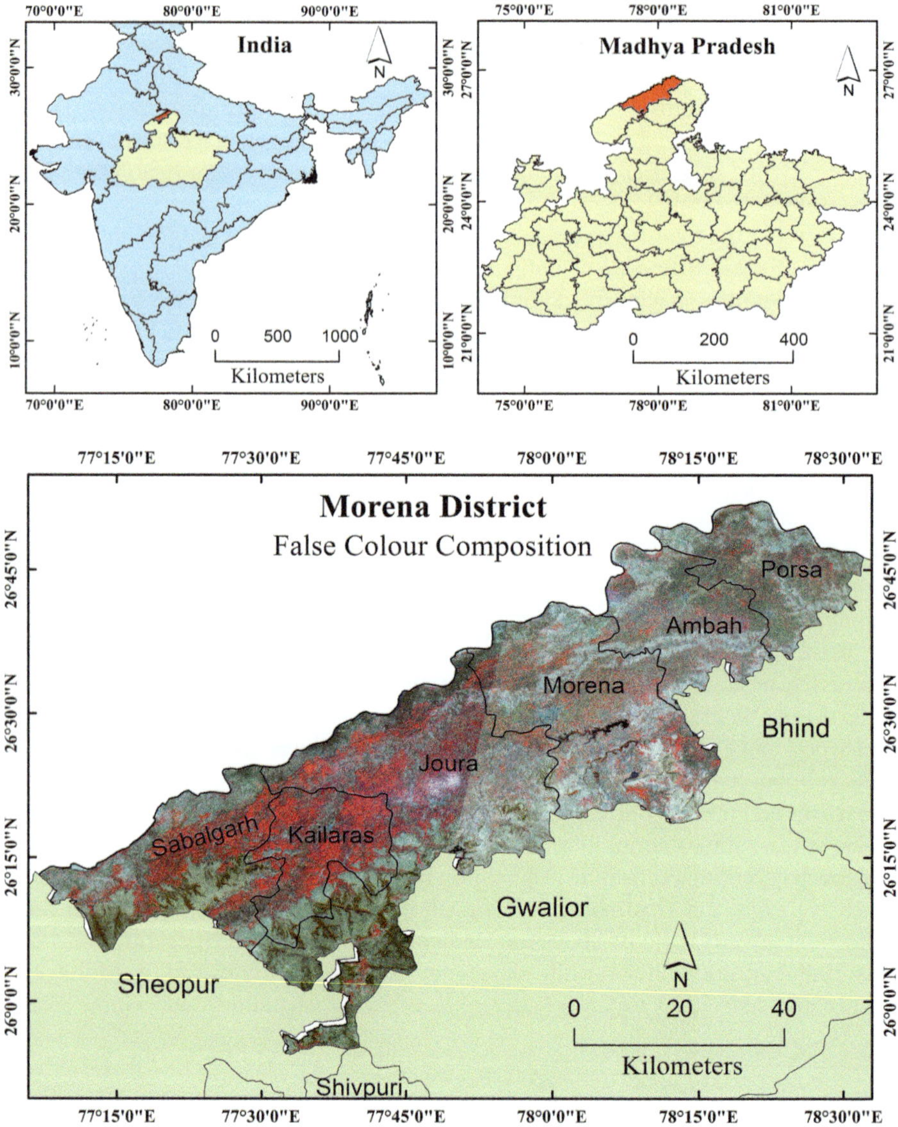

Fig. 2.2 The study area

structurally controlled because of some neotectonic activities that have occurred in different times (Sharma 1980). Chambal is the main river which is flowing from southwest to northeast. Its major tributaries are Kunwari and Asan which drain the area.

The western Marginal Gangetic Alluvial Plain (MGAP) is underlain by recent alluvial accumulated by the recent as well as the ancestral river systems of the Peninsular rivers debouching the Ganga Basin. The basement for the alluvium is formed by the Bundelkhand Granite, the Gwalior Group, and the Vindhyan Supergroup. The Precambrian rocks occur at shallow depth (50–60 m) towards the west in Morena and Dholpur areas, with the sporadic outcrops protruding out of the alluvial cover (as per GSI Report). The major part of the this area is covered by the thick unconsolidated alluvium. This thick alluvium is vulnerable for erosion, and due to erosion the sediment is cut by the small and large gullies which are due course of time connected by each other through headward erosion and form an vast Badlands.

The nature of alluvial deposits of the area is significant not for the agricultural support, but it needs to understand the geomorphic composition to get a clear understanding of the ravine genesis and the stages of ravine development.

2.3.2 Soil

Soils of the lower Chambal valley range widely. In the soil map of India, the lower Chambal valley falls within the limits of two main soil groups, viz. the reddish grey and the yellowish brown alluvial soil. In general, the soils have a surface colour of pale brown to yellowish brown with patches of greyish trace. The soils are broadly of two types: (i) sandy loam to loam with low phosphorous and salt content and (ii) clayey loam with low phosphorous and salt content (GoMP 1996; Pani and Carling 2013). The colour of the soil varies from yellowish brown to dark brown and to dark grey brown in general. The textures of soil in the area are sandy loam, sandy clay loam, clay loam to clay. If we try to understand in detail of the soil nature, we need to combining different scheme of classification the soil of Lower Chambal valley which is suggested major four subsoil types (Sharma 1979). These are as follows:

1.	i. Grey soils without *kankar* layers
	ii. Grey soil with *kankar* layer (below 1 m)
	iii. Grey soil with *kankar* layer (above 1 m)
2.	i. Brown soils without *kankar* layers
	ii. Brown soil with *kankar* layer (below 1 m)
	iii. Brown soil with *kankar* layer (above 1 m)
3.	Reddish soil with well-distributed concentration
4.	Yellowish brown soil with kankar layers below 2 m

Generally the laterite forms plain and occasionally capping over the rocks of Vindhyan Supergroup. Its elevation varies from 400 to 530 m above the m.s.l.. The quaternary alluvium consisting of unconsolidated to consolidated yellowish brown sand silt and clay with gravel and pebbles forms the youngest formation exposed in many parts of the area. The thickness of the alluvium varies from a metre to more than 180 m.

2.3.3 Climatic Conditions

The area falls under the semiarid and the subhumid regions. The climate of Morena district is characterised by a hot summer with dryness apart from Monsoon season (south west). Moderate rainfall, high temperature, dry summer and cold winter are the main features of climate here. The three main seasons are summer (March–June), rainy (July–September) and winter (October–February). Summer is mostly dry and hot, when the mean maximum temperature rises more than 42 °C in May, but the winter is mild, with a mean minimum of 7 °C in January, although it is not uncommon to experience >45 °C during summer and 1–3 °C during winter. The mean annual rainfall varies from 765 mm at Agra to 796 mm at Morena, while Delhi to the north of the area receives 714 mm annually and Gwalior at the south receives 900 mm. Much of the rainfall is received during July–September, and the season for summer monsoon rains, when about 90% of the total annual is received. During the rainy season the relative humidity exceeds 83%. The rest of the year is drier. The driest part of the year is the summer season, when relative humidity is less than 26%. May and early June are the driest months of the year. The wind velocity is higher during the pre-monsoon period. The maximum wind velocity 11.3 km/hr. noted during the month of June and minimum observed in the month of November, e.g. 3.1 km/hr. The average annual wind velocity of Morena district is 6.4 km/hr. There is no meteorological observatory in Morena district. The nearest observatory is Gwalior *(Source: District Ground Water Booklet Copy, Morena District Madhya Pradesh, 2013).*

2.3.4 Vegetation Cover

The entire district is devoid by the vegetation cover. The natural vegetation consists of various kinds of thorny shrubs and trees, including Berry, Dhou, Kardhai, etc. The forest mostly comprises small thorny bushes not consisting of any thickly grown trees, except in some parts of the valleys. However, in some parts of Tonga, Mala and Navez valleys, taller trees of Dhau, Mahua, Jamun, Palas (Tesu), etc. are also present. The ravine land is generally devoid of a good vegetation cover. The predominant species in ravine area are mostly *Anogeissus* pendula. The significant minor forest products are Tendu (*Diospyros melanoxylon*), Khair (*Acacia catechu*) and Harra (*Terminalia chebula*). The ravine area is mostly covered by thorny bushes and trees, and there is scanty vegetative cover. Morena had a forest cover of 14.63%, as against the average of 25.21% for Madhya Pradesh, according to the State of Forest Report 2011 (Forest Survey of India 2011). Moderately dense forest accounts for 13.42%, and open forest accounts for 86.58% of the total forest area, while there is no area under dense forest cover in the district (Forest Survey of India 2011; Pani 2017a).

2.4 Land Degradation in Chambal Valley: Extent

Ravine erosion in India is a very isolated topic in world academic forum. It is still considered as a region-specific problem in India. A very little knowledge had been presented in the world academic forum so far. As per Haigh 1984, the fact that ‘ravine’ erosion, as such, is not recognised elsewhere in the world. The literature on related phenomena, gully erosion, stream trenching and arroyo incision, is heavily

Plate 2.1a The photograph is showing a newly formed gully head erosion more than 2 m depth approaching towards the road

dominated by the work of the American scholars. The region falls under severe ravine erosion category. Nearly 5000 km^2 area of this region is affected by the ravine problem. As per remote sensing data analysed by the researcher, in the year 1974, the total ravine-affected area was 1082.71 km^2 in the study district, which is reduced to 415.98 km^2 in 2014.[5] The total area extension of the ravine visibly cutting down drastically due to the high rate of land levelling all over the ravine affected area in the district. But this cannot be claimed that this practice has been able to check the soil erosion of the study area. Numerous studies have been found that the land levelling accelerates the soil erosion process and the use of levelled land in most of the cases is for agricultural practice which has on-site and off-site implications (Pani 2016a; Marzolff and Pani 2017). During field visit, it has been observed that several new ravine formations are taking place (Plates no 2.1a and 2.1b). This can be taken as a note that still the ravine formation process is active in the study region. The geomorphic process is slow and steady, but the anthropogenic process plays a faster role to subside or modify the natural geomorphological processes. Therefore, the changing ravine environment and area is not indicating the natural ravine formation processes in controlled in the area. The ravine formation processes are still active.

2.5 Causes of Land Degradation

The causes of land degradation especially gully erosion processes have been a widely researched area worldwide, but a clear understanding of gully erosion processes is yet to emerge (Bocco 1991; Foster and Magdoff 1998; Hadley et al. 1985).

[5] Landsat data of 2 years have been used to find the extent and change of ravine area.

Plate 2.1b Plate b is showing how coalition of gully headward erosion engulfed an agricultural land near Bilpur Village in Morena district. (Source: field photograph by author 2015, 2016)

Active gully systems generally develop in unconsolidated soil or debris and frequently result from shifting patterns of land use/land cover and related changes in catchments hydrology (Bocco 1991; Oostwoud, Wijdenes and Bryan 2001). It can be explained as a combination of biophysical processes which played a significant role for occurring land degradation. It is better to state that the gully development chiefly caused by the combination of several processes which may developed simultaneously, by the scouring effect in the bottom and the side walls and sometimes due to hydraulic action of the gully head leading an instant potholes formation which are lately joined and form plunge pool and erode the gully head (FAO 1965).

The overall factors affecting ravine erosion are climate, topography, rock type, vegetation, nature of streams, soil character and the upliftment process which has to be left open for additional factors, such as human interference which may be significant at a particular site.

The problem of land degradation in a form of ravine erosion in Chambal valley is an intense and complex processes. Therefore, the factors are required to be explained in detail.

2.5.1 Natural Factors

2.5.1.1 Geological and Geomorphological Factors

Geological Processes

The Chambal valley is considered to be part of one of the oldest landmasses and landforms which have been formed by the exogenic processes over long geological times. In most parts of this region, gullies are largely more distinct where the rocks are buried by the thick soil or sediment cover. The rocks of the Vindhyan Supergroup are the main source of the soil cover. The shales of the Vindhyan Supergroup are soft in nature and thinly laminated and fissile. There are numerous silty bands. The most important intercalation is of limestones which are stromatolitic at some places. The shales itself are calcareous forming partings at various zones giving rise to *kankars*. The nature and composition of soil is dependent on underlying lithology from which it is originated. The calcareous shales have low permeability, but the presence of *kankar* (calcareous) increases the permeability. The shales normally form gentle slopes because of its easy erodibility. During the monsoon, when the water level is higher, the ravines get flooded, and the water enters into the pores of weathered shales. When the water retracts, blocks of the shales collapse, and thus, there is headward erosion, steep scarps continue, and the valley walls of ravine remain steep. Due to the headward erosion, the side walls also started eroded down.

Slope

Another important factor is slope. Though the slopes do not affect gully initiation, it enhances headward and lateral extension of ravines. The bottom slope of ravine is gentle which depicts a concavity in profile. Slope and soil loss are very closely associated with each other. The force of water erosion is largely a function of slope. The formation of gullies in Chambal ravines is highly affected by slopes.

Topography

The most vital factors affecting the nature and extent of Chambal ravine erosion in the area are formed by topography which is a combination of surface slope, surface length and surface unevenness. The peninsular upland is subject to erosion and degradation by geomorphic processes (Ahmad 1968). Sharma (1968) has noticed that the three erosion surfaces of different geological period range from pre-Cretaceous to the Pleistocene in this Chambal region. It has also been observed in the Chambal River that there are several fluvial originated landforms like river terraces, incised meanders (near Etawah), etc. It has been evident that the region is going through the upliftment processes which is probably the cause for the rejuvenation of Chambal and its tributaries. Pramila Kumar and Rai (1981) also supported the theory of rejuvenation in the central highland. There are numerous geomorphic and fluvial evidences that indicate that the region is undergoing upliftment, and as a result of it, the Chambal and its tributaries have been rejuvenated. Broadly we can say India is under the influence of Himalayan orogeny and the last phase of the Himalayan uplift has an impact on peninsular rivers in India. This factor can be considered as one of the prime factors in ravine genesis in the Yamuna-Chambal region.

2.5.1.2 Climatic Factor

If we consider the climatic factor, the climate, mainly rainfall has a role of promoting a runoff situation in an area and cause for erosion. The erosion caused by rain is determined by the amount, intensity and duration of the rainfall. In the area the maximum intensity of rainfall is received within a span of 3 months. The average annual rainfall of the district is 753.7 mm (CGWB). On the contrary, the remaining months receive no rainfall or minimum rainfall. The erosion processes in the Chambal catchment are further heightened by the regional climate, characterised by irregular and often extremely high rainfall. The rainfall data are scant for the study area. Heavy rainfall frequently results in severe erosion and high rates of soil loss in the Chambal region. The diurnal range of temperature is high in the Chambal region area as it falls in the semiarid climate. This temperature affects runoff by contributing to changes in soil moisture, causing cracks on the land surface during pre-monsoon.

Floods usually do not occur very frequently in the study area. But plain areas along the Chambal are relatively affected by flood hazard.

2.5.1.3 Soil Factor

Soil is the major factor contributes to create network of gullies or formation of Chambal Badlands. The soils of the study area are usually sandy loam to loamy in texture. Nearly one third of the study district area is dominated by thick alluvial soils. It has been witnessed that the soil of this area is comprised with 'kankars' at places and the percentage increases towards the subsurface. With the climate and physiographic anomalies in local environment, the soils become more susceptible to ravine and gully erosion in Chambal.

2.5.2 *Human-Induced Factors*

The biophysical processes regulate soil erosion in numerous way by economic, social and political causes. The human-induced factors have been proved to be one of the major causes of land degradation, soil erosion and biodiversity, as well as threatening food security. The social and economic reasons accelerate the rate of erosion adopted because of unscientific agricultural practices, illiteracy and lack of information for proper land use management practice. More over population dynamics are considered as one of the prime underlying causes of land degradation in this district. The total population growth rate has been increased to 23.44 percentage from 2001 to 2011 (details of socio-economic profiles of various aspects at tehsil level have been analysed and discussed in Chap. 3). However, the direction of the impact depends on the conditions of land use practices.

Increasing abandoned agricultural land because of poverty, lack of manpower due to high rate of migration, poor health and malnutrition, political instability, high demographic pressure and a strange sociological process which are responsible to change the land use practice rapidly of the region which are certainly a key factor for the soil erosion of this region. The GLASOD (Global Assessment of Human-Induced Soil Degradation) survey characterises the following types of contributing factors towards land degradation which are known as human-induced soil erosion.

These are as follows: deforestation and removal of natural vegetation, overgrazing, agricultural activities and industrial activities. Apart from industrial activities, the rest of the above said factors are the major contributing factor of soil erosion of the study area.

2.5.2.1 Vegetation Cover and Grazing Practice

The vegetation cover protects the surface from the impact of raindrop, reduces the amount of water available for runoff by consuming it and by improving infiltration capacity and decreases the velocity of runoff. Thus, vegetation tends to reduce both runoff and erosion. As per the 40 years of land use, change map for the study area is showing there is no dense vegetation cover. In 1974 there was 52.49 percentage of sparse vegetation cover; that too also declined by 52.42 percentage from 1974 to 2014. It has been noticed from the analysed decades of land use data by the author that the sparse vegetation covers are present mainly on the shallow ravine area and the on the gully beds. The ravine lands in the study area are generally devoid of good vegetation. Deforested hills and scanty vegetation areas accelerate the ravine development either on pediment region or on low land areas. In the Chambal ravines, Karonda species, Parbati babul, etc. are the main shrubs and are indicative of the low moisture regime prevailing for most of the year. These plants too also decrease rapidly due to the use of household consumption as a form of fuel wood. Moreover, the high rate of land levelling also major cause of decline the natural vegetation cover of the study area. The field survey indicates that the ravines are developed on low lands where the vegetation is scanty. The ravine lands are considered to be a common land (CPR) which are most of the times used for grazing purposes where animals move freely which in some ways give rise to rill or gully formation. The share of vegetation in villages are less. During the monsoon, torrential downpour through the crop less fields, the current fallows, and the fallow lands is subject to erosion. The intensity of erosion being much higher in the late, which are very much susceptible for ravine erosion and soil loss.

2.5.2.2 Rapid Change of Land Use Practices and Mass Scale Land Levelling

Land use analysis shows that the natural land cover is being modified by human at an increasing rate. Land use changes have amplified considerably in the last few decades. The unsustainable land use practices are a major environmental concern in many developing countries. Land degradation, especially the natural landscape alteration, remains a major ecological problem in this region. Land use management strategies and ravine land conservation strategies have had a significant influence on soil loss in the past 40 years. The increasing rate of built-up land (from 0.19 percentage to 0.68 percentage during 1974–2014) has been observed in the last four decades.[6] It has been observed that the barren land has not been changed substantially, but the percentage of ravine has decreased (from 21.72 percentage to only 8.35 percentage) substantially. Interestingly, the reclaimed ravine land data analysis reveals that the total percentage of reclaimed land have increased at an exponential rate, which are mainly the ravine lands which are levelled by the local farmers for agricultural practice. But it has been observed that agricultural practice on those reclaimed lands is not sustainable and the crop performance is not enough to manage the levelled land for longer time by the farmers. All such reclaimed level land needs continuous monitoring and conservation. The local farmers cannot afford to conserve the land for longer time. Therefore, topsoil loss in reclaimed land is an evitable part of land degradation in this region, and there are increasing rates of off-site implications too (Plate 2.5).

2.5.2.3 Encroachment and Loss of Common Property Resources, Gully Areas and Gully Beds

In the study district, the current development is largely one of leading practice is level the rugged gully land for agricultural practices for to overexploitation in large areas, owing to the lack of alternatives for a growing population. The analysed land use data (Table 2.6) from four decades shows that there is no such significant dense forest cover in the study area present ever. Only 19.54 percentage of sparse vegetation cover is present in the study area; that too also in most of the cases have been noticed that sparse vegetation cover is contributed by the stable ravine land. Due to gully bed widening and gully land levelling, all the natural sparse forest covers are vanishing from this area (Plate 2.3). Therefore, the newly levelled lands are vulnerable during rainy season and are contributing to huge amount of soil in the nearby water body (Plate 2.5). The land levelling trend of this study area has a serious implication for future.

2.5.2.4 Increasing Use of Machinery for Ravine Reclamation

Increasing use of heavy machinery such as JCB, tractor, bulldozer, etc. for land levelling and compacting the ground is a common practice in this region. During the rainy season, the movement of these machines creates artificial small rills which gradually develop a full-fledged gully in and around the villages and agricultural farms (also see Chap. 5).

[6]Apart from the usual increase in built-up area, there are specific reasons for increase in the built-up area in the study region. Because of increasing loss of habitat, wild animals often destroy the standing crops. To save the crops from wild or stray animals, farmers build small houses near their farms, and over time this leads to development of small clusters, called *pura* in the local dialects. The ravine-induced shifting of households and villages also leads to such multiplication of villages/clusters/hamlets.

2.5.2.5 Irrigation and Land Degradation

Most of the agricultural land in the study region are rainfed, although there are pockets of irrigated area. In such areas there is a greater drive for land reclamation. In some of the places, nearer to the canal area like Mrigpura, Ratanbasai, Bagchini, etc., where irrigation facilities are available since the late 1970s, land levelling has increased. But due to the nature of soil, irrigation is not sustainable in this region for a long time (Plates 2.2a and 2.2b).

2.5.2.6 Limited Livelihood Options

The socio-economic data analysis and household survey analysis also reveal that apart from

agriculture (81 percentage people dependent on cultivation, Table 5.2), there are hardly any other livelihood options. Hence, therefore the major pressure is on land. Due to the lack of other alternative livelihood sources in this remote inaccessible region, the resultant pressure on the available land has aggravated land degradation (detail discussed in Chap. 5).

Plate 2.2a Ongoing land levelling on deep ravine. Ridgetop has been removed and creating terrace for soil accumulation at the base of the ravine, Devgarh, Ambah Tehsil. (Source: field photographs from author's field work)

Plate 2.2b Irrigation initiates gully formation in a wheat field levelled 30 years back: Bagchini, Ambah Tehsil

2.5.2.7 Unscientific Agricultural Practice

The unscientific agricultural practices which are followed by the farmers such as farming on the slope without doing any terracing, or modifying the slope to prevent further erosion, also lead to soil erosion. Agriculture on such land on slopes is a common practice in some villages such as Kishroli, Bilpur, Esah and Devgarh.

2.6 Types of Ravines

Ravines are classified in three broad types as per their shape size and intensity: deep ravine, moderately deep ravine and shallow ravine. The described three types of ravines exist together and form an extensive network of ravines. Some of the residue part of the ravine areas also exist within these Badlands which are by nature almost levelled and unconsolidated. These are sometimes developed naturally due to the collapsing or retreating of the side walls, and some of them are levelled mechanically, but due to the nature of the formation, it is very difficult to differentiate the exact nature of extinction of these ravines. However the extinct or relic ravines are categorised as reclaimed or abolished ravine. The classification of ravines has been derived from the basis of severity of the ravine erosion intensity, its shape, size, pattern and cultivation practices. In terms of depth and width and the said parameters, there are three distinct kinds of ravine present in the study area. These are deep ravines which are more than 30 m deep and 2–3 m narrow base width comprised of high steep slope; these are V-shaped. The other type is classified as moderately deep ravine. These are moderately affected by erosion. The depth varies from 5 to 30 m, with base width of 8–20 m; the slope varies from gentle to moderate. These are more of a U-shaped valley. The third one is shallow ravine 1–2 m of depth. There are some areas which are levelled which can be classified as abolished ravines or reclaimed ravine. It has been observed that if base widths are broad, then this area accumulated by thick alluvium transported from the top of the ravine (Fig. 2.3). Often such areas are suitable for agriculture. Interestingly all type of ravines exist together in a form of network of gullies. The above described classifications have been based on the field observations with the help of remote sensing data. The ravine classifications have been attempted by many earlier researchers. The commonly used classifications are as follows.

Based on multiple criteria of shape, size, depth, side slope and head slope into three different types of ravines, such as *young* (narrow base width with steep side slopes), *mature* (wide base width with moderate side slopes) and *old* (very wide base width with gentle side slopes) have been categorised by Pani and Mohapatra (2001) and Pani (2016a) for the lower Chambal valley. The ravine formations in the region are dynamic in terms of shape, size, pattern and aerial extent over the time. The modern geospatial tools like remote sensing and GIS applications. Multi-temporal Landsat images have been successfully used for the mapping of eroded lands. The types of ravines with respect to their average depth into shallow, moderately deep and deep ravines (all qualitative) may be discriminated to some extent based on tone, texture, shape, size and pattern based on elements of remote sensing interpretation techniques. An effort has been made to classify ravines in the study area based on field survey into (i) shallow, (ii) moderately deep and (iii) deep ravines based on their average depth < 5 m, 5 m–20 m and > 20 m, respectively (Tables 2.1, 2.2, 2.3 and Fig. 2.4).

As per the remote sensing decadal image analysis, it has been found that the total ravine area has decreased from 1974 to 2014, with 21.72%–8.35%, respectively.

However, ravine formation processes in the area are still active (Plates 2.1a and 2.1b). The ravines of the study area are rapidly encroaching upon the agricultural land, roads and settlements during the rainy season every year.

2.7 Temporal Changes in Land Use and Land Cover: 1974–2014

The rapid change of land use practice is one of the key factors of land degradation in the study region. The land use of the area mainly encompasses abundant alluvial plains and ravines, and some parts of the district are occupied by barren lands which exist in a scattered manner mainly in south-western parts of the investigated area (Figs. 2.5 and 2.6). The barren lands are basically structural and residual hills which

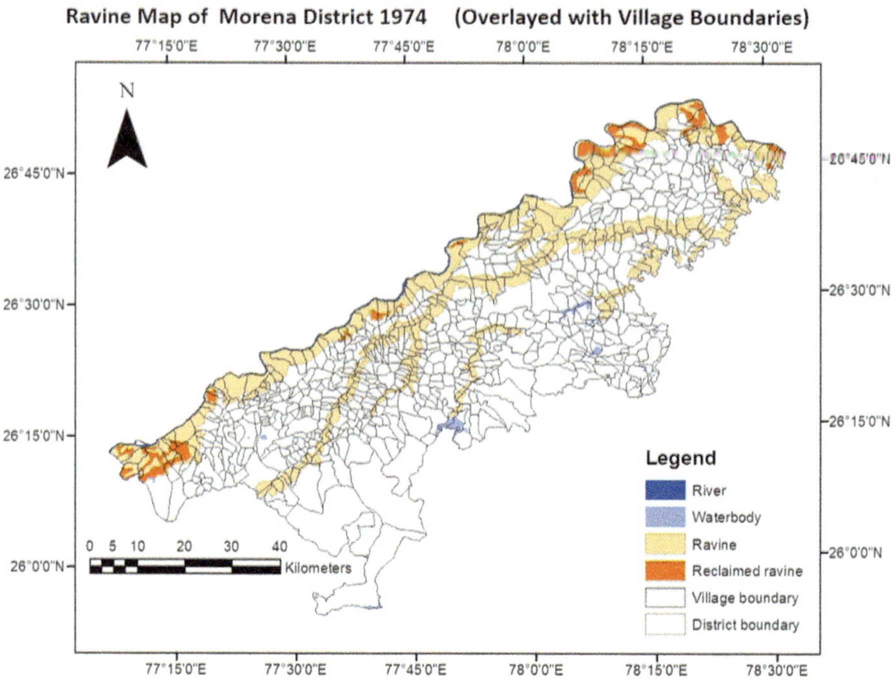

Fig. 2.3 Ravine extension in 1974 in Morena district. (Source: prepared by author based on Landsat 7 (MSS), 1974)

are part of Vindhyan systems. Apart from the southern part of the Sabalgarh area, Pahargarh, southern part of Joura, near Nurabad comes under dissected plateau and residual hills. The structural features observed in the area are in the form of lineaments which are described in the map as barren land (Figs. 2.5 and 2.6). The percentage of barren land is 1.57 which have not been changed so much due to the nature of the barrenness. However, the stone quarrying activities are reported which lead to other problems but did not change the share of barren land of the study region in the last few decades. Interestingly all the barren lands are classified as forest land by different government agencies which are very scanty and sparse category in reality. The other common land use/land covers are mainly rivers, sand bars and water bodies. Small agricultural land is found even in some parts of ravines and inside the ravines and gully beds. The vast area of in this region is comprised of highly dissected ravine and the most striking geomorphic unit of the area (Figs. 2.5 and 2.6). They extend on both flank of the Chambal River and its tributaries, and river Kunwari is one of them. Ravines are characterised by a very fine drainage network and steep head and side slopes with narrow gully divides. Nearly 2–2.5 km area in the both bank of the river Chambal and its tributary ravine has been formed in the region. Interestingly the ravine part is covered by the sparse vegetation cover but rest of the area there is no such vegetation cover presence have been marked. The major part of the area mainly doab area which is a thick alluvium bed between

Table 2.1 Classification of ravines based on morphology

Cites/ type	Gujrat (Tejwani 1974)	Madhya Pradesh (Gupta and Prajapati 1983)	Rajasthan (Bhulyan 1967)	Chambal Region	
				Cited in Haig 1984; Seth et al. 1969)	(Pani and Mohapatra 2001; Pani and Carling 2013; Pani 2016a, 2017a)[a]
Very small	G1: Gullies up to 1 m deep and 18 m bed width	D1: Shallow ravines up to 1.5 m deep	G1: Gullies up to 1 m (18% overall)	D0: Gullies less than 1.5 m deep and 3 m wide (10% overall)	It has been observed that gullies are in general larger than other parts of gullies in India; therefore no such classifications have been attempted
Small	G2: Gullies up to 3 m deep and more than 18 m bed width	D2: Medium ravines 1.5–5 m deep	G2: Gullies, 1–5 m deep (39% overall)	D1: Gullies less than 1.5 m and more than 3 m wide	Shallow ravine: up to 5 m deep and 5–20 m bed base width with gentle to moderately gentle slope
Medium	G3: Gullies 3.9 m deep and not less than 18 m of bed width, side slope 8–15%	D3: Deep ravines, 5–10 m deep	G3: Gullies more than 5 m deep (43% overall)	D3: Trenches 5–10 m deep (6% overall)	Medium ravine: 5–20 m deep, 10–30 m base width with moderate to steep slopes
Large	G4: (a) gullies 3–9 m deep and less than 18 m of bed width (b) Ravines, more than 9 m deep. Side slopes steep to vertical and gullied	D4: Very deep ravines, 10 m deep			Deep ravine: more than 20 m deep; in some places it is more than 60 m deep and has narrow to very narrow base width with steep slopes

[a]For further details, see Pani (2016b)
Source: Ravine classifications of the study region have been done on the basis of the author's field observations (2015, 2016a), and ravine classification of other areas has been compiled from earlier research by various scholars

Table 2.2 Detailed classification of the Chambal ravines (after Sharma 1968)

Particulars of ravine	Description of symbols of ravines		
	G1	G2	G3
Depth in metres	Up to 1	1–5	5 up to 40
Bed width in metres	Up to 18	18–25	Above 25
Slope of head scarp	Gently sloping	Varies	Steep
Slope of sub-scarps	45–80°	50–90°	50–90° But mostly vertical

Source: Sharma (1968)

Table 2.3 Statistics of Badlands area and its change in Morena district year 1974 and 2014

Classes	1974 Area given in km^2	Area percentage	2014 Area given in km^2	Area (percentage)
Ravine	1082.71	88.15	415.98	34.75
Abolished ravine	145.48	11.85	781.09	65.25
Total Badlands area	1228.19		1197.07	

Source: Prepared by author based on Landsat 7 (MSS) (1974) and Landsat 8 (OLI) (2014)

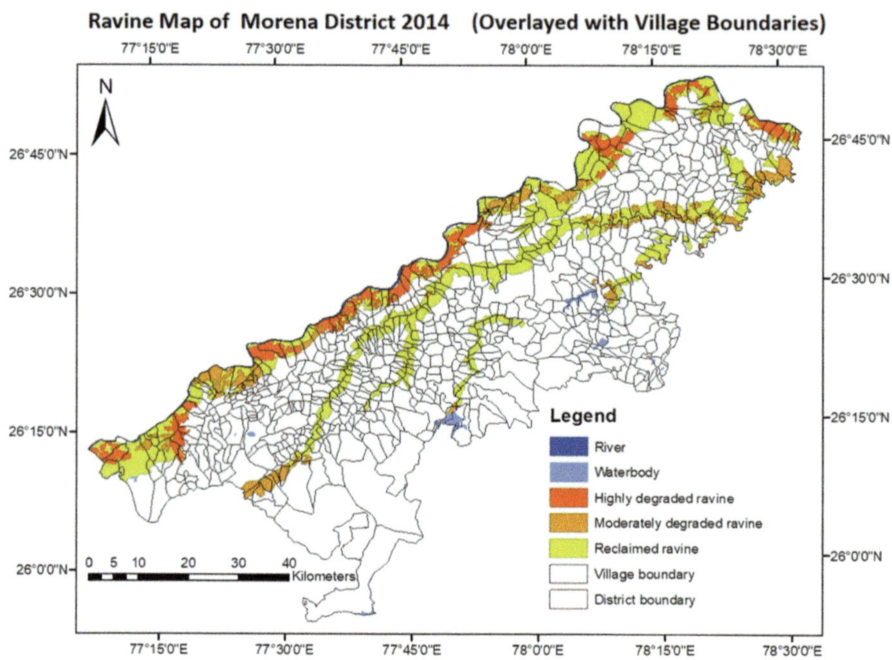

Fig. 2.4 Ravine extension and its classification of Morena district in 2014. (Source: prepared from Landsat 8, 2014)

Chambal and Kunwari is used for the agricultural land. Apart from it, the substantial area has been reclaimed since 1974. Around 2.92% ravine area had been reclaimed during 1974 and 15.67% in 2014 (Table 2.4); subsequently the total ravine area from 1082.71 km^2 in 1974 has been reduced to 415.98 km^2 (Table 2.3). The reclamation process in this ravine is mainly the attempt to level the rugged gully and ravine land for cultivation. It is a common practice since the early 1970s using manual labour, but it has been manifolded in nearly one decade due to the availability of machineries, increasing accessibilities of road (by improved road infrastructure and other governmental development programmes), improving financial means of a section of population and more insistent methods of levelling with bulldozers which have been commissioned during recent years. This results in levelling even observed of the

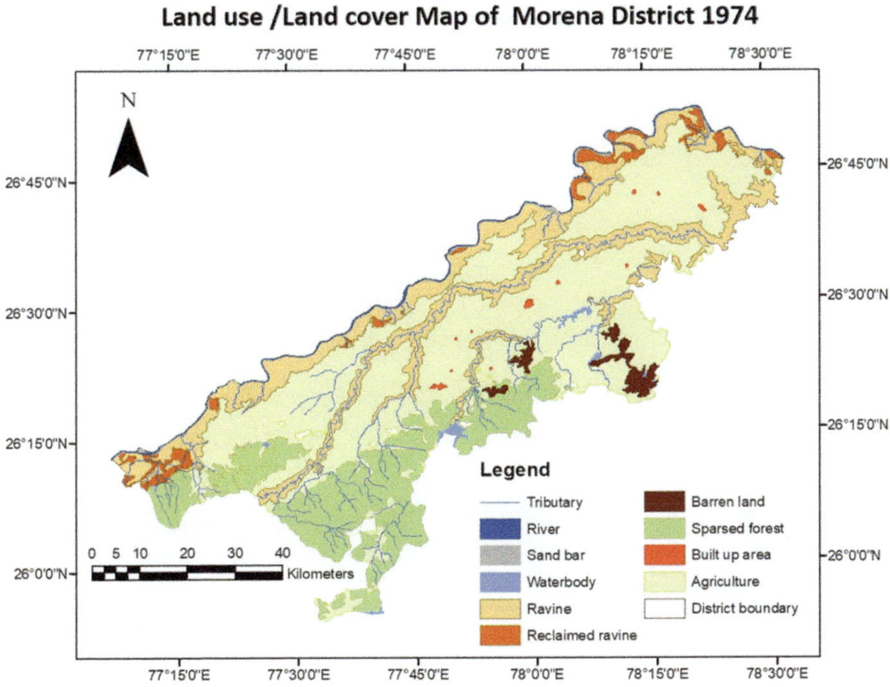

Fig. 2.5 Land use/land cover map of the Morena district (1974)

deeper Badlands (villages Esah, Devgarh, Mrigpura, etc. are examples of deep ravine-affected area) by successive removal of ridgetops, over steepening of side slopes and infilling of ravine bottoms the land levelling activities are manifold over the years.

In Chambal region, there is a long history of Badlands reclamation activities, but very little is known about their actual spatial extent, patterns and temporal dynamics. Official wasteland reclamation statistics usually do not differentiate Badlands from other wasteland, nor do they include non-governmental reclamation projects conducted by individual farmers or village communities (Marzolf and Pani 2017). With increasing population pressure and agricultural mechanisation, however, we can expect and already observe in the field an increasing encroachment into the Badlands. Usually the shallower badlands adjacent to agricultural areas on the alluvial plains are smoothed by manual work and with tractors or bulldozers into slightly undulating terrain. It has been observed that in the last few years, the deeper badlands are also getting levelled by the locals using heavy machinery, such as bulldozer, tractor, etc. In most of the cases, they follow the terrace levelling methods for the deep ravine to reduce the cost and timing of levelling. The ridgetops of deeper badlands are removed initially, and the ridgetop materials are used for infilling ravine bottoms. More than 40-m-deep ravine is also getting levelled in several places. Villages Deogarh, Esah, Mrigpura and Kishroli are among them. An

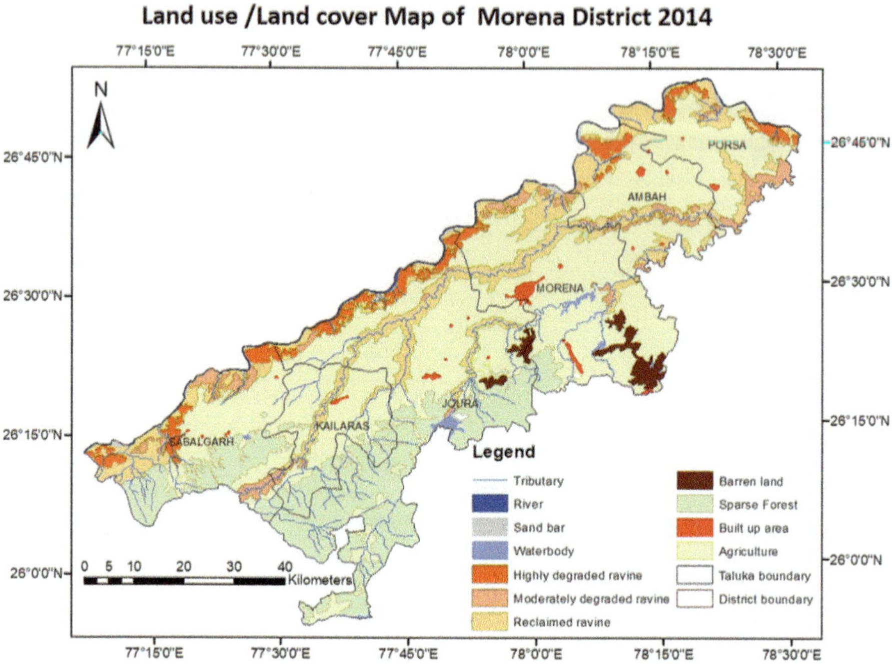

Fig. 2.6 Land use/land cover map of Morena district 2014

Table 2.4 Land use/land cover statistics of Morena district 1974

Land use classes	1974	Percentages
Built up	9.24	0.19
Agriculture	2615.79	52.49
Sparse Forest	961.11	19.28
Ravine	1082.71	21.72
Reclaimed ravine	145.48	2.92
Barren land	78.29	1.57
Waterbody	24.74	0.50
Sandbar	32.98	0.66
River	33.49	0.67
Total	4983.83	

Note: Area given in km^2

example for deep-badland levelling is shown in Plate 2.2a, where a sub-catchment of the main ravine system was completely re-modelled. The outlet of the sub-catchment was closed with an earthen dam drained by a concrete pipe. The topography of the 9000 m^2 area is uneven with lower parts remaining in the old valley positions; farmers expect eroded sediment to accumulate at the base of the ravine and to increase soil thickness and fertility over the next few years. Newly levelled land is commonly left uncultivated at the beginning for it to compact and to increase soil moisture first rainy season after levelling. Afterwards, mustard and millet are sown as first crops. However, agricultural use that often use for food crop to fulfil the need of a household. Field observations, satellite images analysis and statements by local farmers suggest enhanced erosion rates on reclaimed land as compared to original undisturbed agricultural fields. The sustainability of such reclaimed land is questionable and required unbiased research to understand the dynamics completely in the region (Marzolf and Pani 2017).

Most of the levelled ravine land have been used as an agricultural land in the district (implications discussed in Chap. 5 while discussing village-level household survey). Thus, the areal contraction of ravines has been in the magnitude of around 8.35% over the last 40 years (1974–2014) (Tables 2.5 and 2.6). So far as changes in other land use classes are concerned, the major changes are an increase in built-up area and agricultural land and the decline in forest cover; however the forest cover was always sparse in nature in the study district.

Forty years of remote sensing data analysis by the author brought enough evidences of change of land use practice in the study region. Loss of river water, change of natural topography and acceleration of the levelled ravine land (12.75 percentage increased during 4 decades) are shown to be the trends of rapid change of land use/ land cover of the study region. Natural gully formation accelerated by the intensification of farming systems (Plates 2.1a and 2.1b). The depletion of the soil organic matter, the instable soil structure and levelling of the deep ravine intensify the soil loss and increase by the runoff process. It has been observed that apart from loosening the soil by levelling, the levelled land, when irrigated, may result in gully erosion. Sometimes, gully initiation takes place under similar circumstances. During field work, it has been observed that the levelled ravine in Bagchini village on the bank of Kunwari river, which has a long history of mass scale land levelling, are also experiencing gully formation while irrigating the levelled land (field plates 2.2a

Table 2.5 Land use/land cover statistics of Morena district 2014

Land use classes	2014	Percentage
Built up	33.70	0.68
Agriculture	2612.14	52.42
Sparse forest	973.64	19.54
Ravine	415.98	8.35
Reclaimed ravine	781.09	15.67
Barren land	78.29	1.57
Waterbody	24.75	0.50
Sandbar	31.63	0.63
River	32.14	0.64
Total	4983.35	

Note: Area given in km^2
Source: Computed by the author from the remote sensing (Landsat 8, 2014) data

Table 2.6 Change in land use/land cover statistics of Morena district 1974–2014

Land use classes	1974	2014	Percentages 1974	Percentages 2014	Change 1974–2014
Built up	9.24	33.70	0.19	0.68	0.49
Agriculture	2615.79	2612.14	52.49	52.42	–0.07
Sparse Forest	961.11	973.64	19.28	19.54	0.26
Ravine	1082.71	415.98	21.72	8.35	−13.37
Reclaimed Ravine	145.48	781.09	2.92	15.67	12.75
Barren land	78.29	78.29	1.57	1.57	0
Waterbody	24.74	24.75	0.50	0.50	0
Sandbar	32.98	31.63	0.66	0.63	–0.03
River	33.49	32.14	0.67	0.64	–0.03
Total	4983.83	4983.35			

Note: Area given in km^2

Source: Computed by the author from the remote sensing (Landsat (MSS), 1974, and Landsat, 8 2014) data of 40 years

Plate 2.2c In the same field a closer view of the gully formation. (Source: field photographs from author's field work)

and 2.2b). These plates also show gully initiation due to irrigation in the levelled land of village Bagchini, which were created by demolishing ravines (Reference Plates 2.2b and 2.2c). Other researchers from different regions across the globe also report that irrigation can also result in gully erosion (e.g. Vanacker et al. 2003; Nyssen et al. 2004; Poesen et al. 2003).

The land use/land cover investigation is based on the reason that skewed allocation of farmland has caused unsustainable land use behaviour that is creating land degradation, although there are many dimensions to land degradation. There are

Plate 2.3 Massive land levelling is going on in Khurd Village, Ambah Tehsil, on the bank of Kunwari River, a major tributary of Chambal. (Source: field photograph from author's field work)

several examples in the field which can be cited as instant examples of the above statement. The ravine lands are invariably levelled down for agriculture practices for the last few years in a rapid manner without following any scientific approach (Plate 2.3). These levelled lands are the most vulnerable towards soil erosion, and it is difficult to check the flow during the rainy season due to high runoff. Created bunds to check the erosion are also not sustainable in such levelled lands (Plates 2.4a and 2.4b). There are many off-site impacts of land degradation, as there is huge siltation in the nearby water bodies due to high soil erosion (Plate 2.5).

2.8 Conclusion

There are various causes of land degradation, and some of which, such as the geologic factors, are beyond human control. Therefore, gully and ravine formation, to some extent, is inevitable in this region. But other factors like geomorphology, slope and the human-induced factors can be minimised and managed to reduce ravine formation. It is possible with suitable agricultural practice, proper land management and sustainable land reclamation policy. But the rapid change of land use practice, discussed in the chapter, is one of the key factors causing land degradation. Utilising available data from different sources, such as remote sensing data, published evidences or information on land use and a deeper understanding of the nature of land degradation in the region would certainly help to minimise land degradation of the

Plate 2.4a Series of small earthen bunds constructed to check runoff and soil erosion on the recently reclaimed levelled ravine land for agriculture (near village Esah, Morena)

Plate 2.4b Earthen bund constructed on the reclaimed ravine land is no longer sustainable, a closer view of an eroded bund (near village Esah, in Morena). (Source: field photograph from author's field work)

Plate 2.5 Off-site impact, huge amount of siltation in the river Kunwari, near Bagchini village, due to land levelling practices and the soil loss of the area. (Source: field photograph from author's field work)

region. Unequal access to land resources creates a social, economic and environmental threat to the people of the region (Pani 2017a, b). Erosion and unscientific land use practices, mainly large-scale land levelling, not only deteriorate the soil quality on-site but also result in significant sediment-related problems off-site.[7]

References

Ahmad E (1968) Distribution and causes of gully erosion in India. Selected papers Seminar, p 21

Bhulyan S (1967) Survey of ravine lands in Rajasthan. In: Proceedings of the 11th Silvi-culture conference. Forest Research Institute, Dehradun, UP (now Uttarakhand)

Bocco G (1991) Gully erosion: processes and models. Prog Phys Geogr 15(4):392–406

Chaudhuri S, Gupta N (2009) Levels of living and poverty patterns: a district-wise analysis for India. Econ Polit Wkly:94–110

Das DC (1985a) Problem of soil erosion and land degradation in India. In: Proceedings of the National seminar on soil conservation and watershed management, New Delhi

Das DC (1985b, September) Problem of soil erosion and land degradation in India. In: National seminar on soil conservation and watershed management, pp 17–18

FAO (1965) Report the state of food and agriculture. FAO, Rome. www.fao.org/docrep/017/ap653e/ap653e.pdf

Forest Survey of India (2011) India: state of forest report 2011, Dehradun. Available at: http://fsi.nic.in/details.php?pgID=sb_16. Accessed 5 July 2016

[7] In the field photo Plate 2.5, the sedimentation problem due to unchecked soil erosion in the levelled land can be seen, which leads to deposited huge sediment in the river system.

Foster JB, Magdoff F (1998) Liebig, Marx, and the depletion of soil fertility: Relevance for today's agriculture. Mon Rev 50(3):32
Government of Madhya Pradesh (1996) Madhya Pradesh District gazetteers. Government of Madhya Pradesh, Morena/Bhopal
Government of Madhya Pradesh (2007) Madhya Pradesh human development report, 2007 [online]. Accessed 20 May 2011. http://www.mp.gov.in/difmp/mphdr/MPHDR2007_English.pdf
Gupta RK, Prajapati MC (1983) Reclamation and use of ravine lands. Desert Resour Technol (Jodhpur) 1:221–262
Hadley RF, Lal R, Onstad CA, Walling DE, Yair A (1985) Recent developments in erosion and sediment yield studies. IHP-II Project A. 1.3. 1. UNESCO
Kandrika S, Dwivedi RS (2013) Reclamative grouping of ravines using Cartosat-1 PAN stereo data. J Indian Soc Remote Sens 41(3):731–737
Kumar P, Rai RK (1981) Has Deccan foreland rejuvenated. Perspect Geomorphol 1:101
Marzolff I, Pani P (2017) Dynamics and patterns of land levelling for agricultural reclamation of erosional badlands in Chambal Valley (Madhya Pradesh, India). Earth Surf Process Landf. https://doi.org/10.1002/esp.4266
Ministry of Agriculture (1984) Report of the working group on reclamation and development of ravines for formulation of the five year plan, New Delhi
NCA (1976) Report of the National Commission on Agriculture. Government of India, Ministry of Agriculture and Irrigation, New Delhi
Nyssen J, Poesen J, Moeyersons J, Deckers J, Haile M, Lang A (2004) Human impact on the environment in the Ethiopian and Eritrean highlands – a state of the art. Earth Sci Rev 64(3–4):273–320
Oostwoud Wijdenes DJ, Bryan R (2001) Gully-head erosion processes on a semi-arid valley floor in Kenya: a case study into temporal variation and sediment budgeting. Earth Surf Process Landf 26(9):911–933
Pani P (2016a) Controlling gully erosion: an analysis of land reclamation processes in Chambal Valley, India. Dev Pract 26(8):1047–1059
Pani P (2016b) Controlling gully erosion: an analysis of land reclamation processes and challenges in Chambal badlands, India. Geophys Res Abstr 18:765
Pani P (2017a) Ravine erosion and livelihoods in semi-arid India: implications for socioeconomic development. J Asian Afr Stud 53(3):437–454
Pani P (2017b) Chambal without ravines. Down Earth 26(8):80–81
Pani P, Carling P (2013) Land degradation and spatial vulnerabilities: a study of inter-village differences in Chambal Valley, India. Asian Geogr 30(1):65–79. https://doi.org/10.1080/10225706.2012.754775. Routledge
Pani P, Mohapatra SN (2001) Delineation and monitoring of gullied and ravinous lands in a part of lower Chambal Valley, India, using remote sensing and GIS. 22nd Asian conference on remote sensing, vol 54, pp 5–9
Poesen J, Nachtergaele J, Verstraeten G, Valentin C (2003) Gully erosion and environmental change: importance and research needs. Catena 50(2–4):91–133
Sharma HS (1968) Genesis of ravines of the lower Chambal Valley, India. In: Selected papers, 21st International Geographical Union Congress, vol 1, pp 114–118
Sharma HS (1979) The physiography of the lower Chambal Valley and its agricultural development: a study in applied geomorphology. Concept Publishing Company, New Delhi
Sharma HS (1980) Ravine erosion in India. Concept Publishing Company, New Delhi
Tejwani KG (1974) Classification and reclamation of gullied lands. J Soil Water Conserv India 16:24–25
Vanacker V, Govers G, Barros S, Poesen J, Deckers J (2003) The effect of short-term socio-economic and demographic change on landuse dynamics and its corresponding geomorphic response with relation to water erosion in a tropical mountainous catchment, Ecuador. Landsc Ecol 18(1):1–15

Chapter 3
Socio-economic Scenario in a Badlands Region

Abstract The third chapter is focused on the socio-economic profile of the study region, and the fourth one covers the linkages between land degradation, rural development and livelihoods. This chapter contextualises the levels of economic development in the disaggregated administrative units with the overall patterns at the district and the state levels. Further, sub-district level differences point to the specific patterns of socio-economic development in a ravine-affected region. Another insight from the exercise is to understand the overall nature of social differentiations in the extent of development, captured through a limited but significant set of variables.

Keywords Socio-economic profile of Badlands · Demographic profile · Madhya Pradesh · Social groups · Village Directory · Primary Census Abstract

3.1 Introduction

Rural development is a multi-dimensional process. The idea of development itself has expanded from a narrow focus on income and employment to incorporate various dimensions of well-being (Sen 1999). Rural development, as a concept that emphasises the betterment of the rural population, has a spatial element built into it. However, from the perspective of sustainable development, there is a focus on the nature and implications of the development process as well. Thus, rural development encompasses fundamental restructuring of the rural economy (viz. rising productivity in agriculture, diversification of employment and livelihoods, expansion of the rural non-farm sector) and also involves changes in the ways available natural resources are utilised by the people. The processes can be examined at multiple levels, such as at the levels of region, village, households or individual. A crucial component of understanding the process is the demographic and socio-economic characteristics of the population at a disaggregated level. As it has been discussed in the conceptual framework adopted in the study (in Sect. 1.11), population characteristics bring out

P. Pani, *Land Degradation and Socio-Economic Development*, Advances in Asian Human-Environmental Research,
https://doi.org/10.1007/978-3-030-42074-1_3

some of the crucial features of the level of development in the study region. These characteristics help in contextualising the later discussions on land degradation and its linkages with the process of rural development. The broad overview of the socio-economic scenario of the study region, particularly with respect to the sectoral distribution of the workers, helps in outlining the empirical context of the research.

The relationship between the natural environment and economic development depends a lot on the specific regional context. The process of development is shaped through the interaction of human society and nature. While natural resources are used as a source of livelihoods, human activities affect the status of the environment. To understand the implications of land degradation in the context of Chambal valley and particularly that of Morena district, a broad overview of the study region has been presented in this chapter. This analysis is mostly based on secondary data at a disaggregated (tehsil) level. The entire exercise has been carried out to put the different indicators for Morena district in comparison with that of Madhya Pradesh and India. There are six tehsils in the district, and all the districts are not equally affected by land degradation. Therefore, the tehsil or block-wise data has also been presented to highlight the differences within the district. A combined analysis of biophysical and environmental factors suggests significant differences in the extent of land degradation at the tehsil level. The analysis in this chapter also brings out the differences at the tehsil level.

The various parameters through which the patterns of socio-economic development in the region have been investigated include population, population growth, shares of rural and urban populations, gender composition of the population, average family size, sex ratio, child sex ratio, dependency ratio, social and religious composition, literacy rates, the distribution of workers across different industries, etc.

3.2 Population

Population characteristics are important to understand the livelihoods and development scenario of a region. In this section we have discussed about the size of population and population growth.

3.2.1 Madhya Pradesh

Madhya Pradesh is a state in Central India and is also popularly known as the 'Heart of India' due to its geographical location. Before its separation to two states, namely, Madhya Pradesh and Chhattisgarh, in 2000, Madhya Pradesh was the largest state in India. Chhattisgarh was carved out of Madhya Pradesh, and it comprised of seven districts of erstwhile Madhya Pradesh (viz. Surguja, Bilaspur, Raigarh, Rajnandgaon, Durg, Raipur and Bastar). Table 3.1 shows the total population figures of Madhya

Table 3.1 Total population of males and females in Madhya Pradesh in 1991, 2001, 2011

	Population 1991			Population 2001			Population 2011		
Sector	Total	Male	Female	Total	Male	Female	Total	Male	Female
Total	66,181,170	34,267,293	31,913,877	60,348,023	31,443,652	28,904,371	72,626,809	37,612,306	35,014,503
Rural	50,842,333	26,164,353	24,677,980	44,380,878	23,031,093	21,349,785	52,557,404	27,149,388	25,408,016
Urban	15,338,837	8,102,940	7,235,897	15,967,145	8,412,559	7,554,586	20,069,405	10,462,918	9,606,487

Source: Census of India 1991, 2001, 2011

Pradesh over three time periods of 1991, 2001 and 2001, by both gender and sectoral location. However this depicts the unaltered population figures of Madhya Pradesh (including Chhattisgarh total). These population figures of 1991 are noncomparable with that of subsequent later time periods (i.e. 2001 and 2011).

The next table (Table 3.2) shows total population figures of Madhya Pradesh with adjustments in 1991 population (excluding Chhattisgarh in 1991). From Table 3.2 it can be seen that the total population of Madhya Pradesh has increased from 48,566,242 in 1991 to 72,626,809 in 2011. About 52% of the total population of Madhya Pradesh in 1991 were males, and around 47% were Females. This male-female percentage population ratio over 2001 and 2011 broadly remains the same, with minor variations in the rural population. There is a slightly higher share of females in the rural population (as could be seen in Table 3.3).

The rural-urban share of the total population of Madhya Pradesh however fluctuates from 1991 to 2001 to 2011. This can be seen in Table 3.4 where the share of urban population is marginally increasing (from around 25% in 1991 to 27.7% in 2011).

Madhya Pradesh being the second largest state in India (in terms of its area) and fifth largest state in India (based on its population size), there is considerable increase in total population over the two time periods (1991–2001 and 2001–2011). The most recent time period has seen larger increase in absolute population as compared to the earlier (see Table 3.5).

However when considering the population growth rates of Madhya Pradesh during 1991–2001 and 2001–2011, it is found that the state had a higher growth rate of 24.26% in the former period and it had declined to 20.35% in 2001–2011. Population growth rate based on rural-urban location has slight differences; urban areas have higher growth rates as compared to the rural areas. However even these shares have declined over time.[1] Also considering the gender dimension of population growth rates, it is found that males have a slightly lower growth rates as compared to that of females (refer to Table 3.6).

3.2.2 *Morena District*

Morena district is one of the 51 districts of present state of Madhya Pradesh under the Chambal administrative division and is situated in the extreme northern part of the state. It further comprises of six tehsils – Ambah, Porsa, Morena, Joura, Kailaras and Sabalgarh. Morena district before 1998 comprised of eight tehsils. Two tehsils of Morena district – Vijaypur and Sheopur – were separated out from Morena district in 1998 to form a separate district of Sheopur. The total population of Morena district in terms of its absolute population figures has increased from 2001 to 2011 (Table 3.7).

[1] From 1991 to 2001 and 2001 to 2011, the urban population growth rates have declined from 30 per cent to 25.7 per cent, whereas the rural population growth rates have declined from 22.3 per cent to 18.4 per cent.

Table 3.2 Total population of Madhya Pradesh in 1991 (adjusted), 2001, 2011

	Population 1991			Population 2001			Population 2011		
Sector	Total	Male	Female	Total	Male	Female	Total	Male	Female
Total	48,566,242	25,394,673	23,171,569	60,348,023	31,443,652	28,904,371	72,626,809	37,612,306	35,014,503
Rural	36,292,098	18,890,291	17,401,807	44,380,878	23,031,093	21,349,785	52,557,404	27,149,388	25,408,016
Urban	12,274,144	6,504,382	5,769,762	15,967,145	8,412,559	7,554,586	20,069,405	10,462,918	9,606,487

Note: 1991 figures include M.P. minus seven districts that later formed Chhattisgarh (Surguja, Bilaspur, Raigarh, Rajnandgaon, Durg, Raipur, Bastar)
Source: Census of India 1991, 2001, 2011

Table 3.3 Share of male, female and total population in Madhya Pradesh in 1991, 2001, 2011 (rural and urban)

	Population 1991			Population 2001			Population 2011		
Sector	Male	Female	Total	Male	Female	Total	Male	Female	Total
Total (R + U)	52.29	47.71	100	52.10	47.90	100	51.79	48.21	100
Rural	52.05	47.95	100	51.89	48.11	100	51.66	48.34	100
Urban	52.99	47.01	100	52.69	47.31	100	52.13	47.87	100

Source: Census of India 1991, 2001, 2011

Table 3.4 Share of rural and urban population of male, female and total population in Madhya Pradesh in 1991, 2001, 2011

	Population 1991			Population 2001			Population 2011		
Sector	Male	Female	Total	Male	Female	Total	Male	Female	Total
Rural	74.73	74.39	75.10	73.54	73.25	73.86	72.37	72.18	72.56
Urban	25.27	25.61	24.90	26.46	26.75	26.14	27.63	27.82	27.44
Total	100	100	100	100	100	100	100	100	100

Source: Census of India 1991, 2001, 2011

Table 3.5 Increase in population of Madhya Pradesh in 1991–2001 and 2001–2011

	1991–2001 population			2001–2011 population		
	Total	Male	Female	Total	Male	Female
Total	11,781,781	6,048,979	5,732,802	12,278,786	6,168,654	6,110,132
Rural	8,088,780	4,140,802	3,947,978	8,176,526	4,118,295	4,058,231
Urban	3,693,001	1,908,177	1,784,824	4,102,260	2,050,359	2,051,901

Source: Census of India 1991, 2001, 2011

Table 3.6 Population growth rates of Madhya Pradesh in 1991–2001 and 2001–2011

	1991–2001 growth rate (in percentages)			2001–2011 growth rate (in percentages)		
	Total	Male	Female	Total	Male	Female
Total	24.26	23.82	24.74	20.35	19.62	21.14
Rural	22.29	21.92	22.69	18.42	17.88	19.01
Urban	30.09	29.34	30.93	25.69	24.37	27.16

Source: Census of India 1991, 2001, 2011

Morena district has around 54.8% male population and 45.2% female population, which has slightly higher percentage of male population share as compared to the overall state share of 52%. Though the male-female share of Morena district remains almost same over time, sector and gender, nevertheless the rural areas have slightly higher share of male population as compared to the urban population (as can be seen in Table 3.8). From 1991 to 2011, the percentage of rural population has marginally decreased, with a proportionate increase of urban population (as in Table 3.9).

Table 3.7 Total population of Morena district for total, males and females in 1991–2011

	Population 1991			Population 2001			Population 2011		
Sector	Total	Male	Female	Total	Male	Female	Total	Male	Female
Total	2,269,589	1,255,883	1,013,706	1,592,714	874,089	718,625	1,965,970	1,068,417	897,553
Rural	990,495	548,406	442,089	1,249,409	687,664	561,745	1,495,508	815,218	680,290
Urban	1,279,094	707,477	571,617	343,305	186,425	156,880	470,462	253,199	217,263

Source: Census of India 2001, 2011

Table 3.8 Share of male, female and total population in Morena district in 1991, 2001 and 2011 in both sectors

	1991			2001			2011		
Sector	Male	Female	Total	Male	Female	Total	Male	Female	Total
Total (R + U)	54.78	45.22	100	54.88	45.12	100	54.35	45.65	100
Rural	54.78	45.22	100	55.04	44.96	100	54.51	45.49	100
Urban	54.78	45.22	100	54.30	45.70	100	53.82	46.18	100

Note: The figures for 1991 refer to the old Morena district
Source: Census of India 1991, 2001, 2011

Table 3.9 Share of rural and urban population of male, female and total population in Morena district in 1991, 2001, 2011

	Population 1991			Population 2001			Population 2011		
Sector	Male	Female	Total	Male	Female	Total	Male	Female	Total
Rural	79.48	79.48	79.48	78.45	78.67	78.17	76.07	76.30	75.79
Urban	20.52	20.52	20.52	21.55	21.33	21.83	23.93	23.70	24.21
Total	100	100	100	100	100	100	100	100	100

Note: The figures for 1991 refer to the old Morena district
Source: Census of India 1991, 2001, 2011

Morena district is growing at 23.4% which is slightly lower than overall state average population growth rate of 24%. However this population growth rate has massive rural-urban disparity with rural areas growing at lesser rate (19.7%), and urban areas have fast growth rate (37%). Like Madhya Pradesh, Morena district also has a slightly higher population growth rates for its females (Table 3.10).

In terms of basic demographic parameters, Morena district reflects a high growth of population during 2001–2011 period and also a substantially higher growth of population in the urban areas than in the rural area; but the decadal rate of growth of urban population is much higher in the Morena district than in the whole of Madhya Pradesh. However, as the subsequent analysis shows, there are important tehsil-level differences within the district.

3.2.3 *Comparative Percentage Shares of Rural and Urban Populations: M.P., Morena District and Its Six Tehsils*

The comparative assessment of the shares of male and female populations for Madhya Pradesh, Morena district and its tehsils has been presented in Table 3.11. The data presented in Table 3.11 shows that Morena district has a comparatively lower share of females. This is more or less the scenario in the tehsils as well.

The share of rural and urban populations has been presented in Table 3.12. It is found that the district Morena is less urbanised than Madhya Pradesh. The share of urban population is the highest in Morena Tehsil and is the lowest in the Joura Tehsil (Fig. 3.1).

Table 3.10 Population growth rates in Morena district for 2001–2011

	2001–2011 growth rate (in percentages)		
T/R/U	Total	Male	Female
Total	23.44	22.23	24.9
Rural	19.7	18.55	21.1
Urban	37.04	35.82	38.49

Source: Census of India 2001, 2011

Table 3.11 Share of male and female population in 1991, 2001, 2011

Area	M_91	M_2001	M_2011	F_91	F_2001	F_2011
	Males			Females		
Madhya Pradesh	51.78	52.10	51.79	48.22	47.90	48.21
Morena district	54.78	54.88	54.35	45.22	45.12	45.65
Ambah Tehsil	55.18	54.69	53.88	44.82	45.31	46.12
Poura Tehsil	54.65	54.06	53.71	45.35	45.94	46.29
Morena Tehsil	55.86	55.39	54.84	44.14	44.61	45.16
Joura Tehsil	55.61	55.30	54.61	44.39	44.70	45.39
Kailaras Tehsil	54.89	54.60	54.09	45.11	45.40	45.91
Sabalgarh Tehsil	54.52	54.09	53.87	45.48	45.91	46.13

Source: Census of India 1991, 2001, 2011

Table 3.12 Percentage of rural population in 1991, 2001, 2011

Rural Area	T_91	T_2001	T_2011
Madhya Pradesh	77.33	73.86	72.56
Morena district	79.48	78.17	75.79
Ambah Tehsil	85.10	83.25	81.49
Poura Tehsil	86.30	83.33	82.51
Morena Tehsil	54.51	62.53	56.86
Joura Tehsil	86.96	90.43	90.26
Kailaras Tehsil	100	85.76	86.38
Sabalgarh Tehsil	77.75	76.80	77.02

Source: Census of India 1991, 2001, 2011

3.3 Demographic Profile of Madhya Pradesh, Morena District and Its Six Tehsils

The demographic profiles of the following spatial units (Madhya Pradesh, Morena district, six tehsils of Morena district – Ambah, Porsa, Morena, Joura, Kailaras and Sabalgarh) are examined with respect to the following indicators over the three time periods of 1991, 2001 and 2011: average family size, sex ratio, child sex ratio, child-woman ratio, dependency ratio, young and aged dependency ratio.

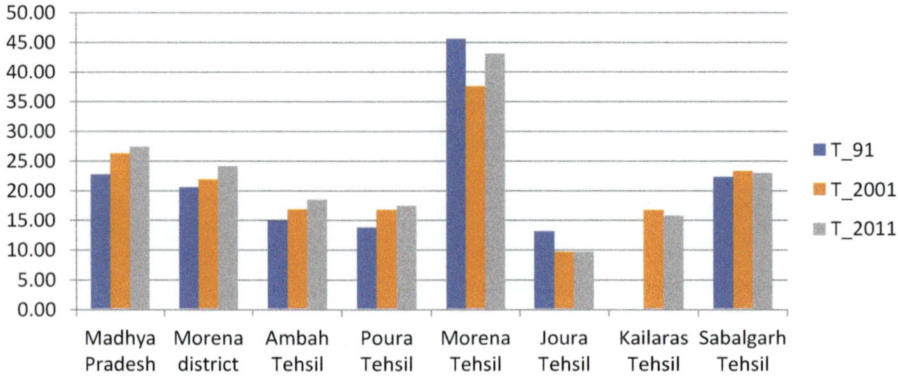

Fig. 3.1 Urban share (in percentage) of total population in 1991–2011 for M.P., Morena district and its six tehsils

3.3.1 Madhya Pradesh

The sex ratio of Madhya Pradesh in 1991 was 912 which were lower than national average; its child sex ratio was somewhat higher (941). The average family size was around 5.8 (Table 3.13). There are prominent rural-urban differences with the sex ratio of urban areas being extremely low (887). Also, the family size of rural areas in Madhya Pradesh is somewhat higher than that of its urban areas (5.85 and 5.65, respectively).

The average family size somewhat declined to 5.53 in 2001 from 5.8 in 1991. The sectoral gap also declined (5.55 in rural areas and 5.48 in urban areas). The sex ratio slightly improved from 912 in 1991 to 919 in 2001; although it was still lower than the national average. Urban Madhya Pradesh's sex ratio is still quite low – 898. Further, the child sex ratio alarmingly declined in both rural and urban areas, especially in urban areas, from 931 in 1991 to around 907 in 2001. The child-woman ratio was 531 in 2001, with it being higher still in rural areas (586) than the urban areas (398). Urban women seem to have lesser children. Considering the dependency ratio, Madhya Pradesh has 84% dependency ratio. This is higher still for its rural areas (nearly 91%) and somewhat low for its urban areas (68%). Madhya Pradesh has a high young dependency ratio (971%), even higher for its rural areas (77%) (Table 3.14).

The decline in average family size continued from 1991 to 2001 to even 2011, with it going down to 4.8. However, unlike earlier time periods, the family size of urban areas increased to five per family. Sex ratio again increased during 2001–2011 to 930, somewhat lower than the national average. Urban sex ratio is still lower at 918 in 2011 (Table 3.15). Decline in child sex ratio continued from 2001 to 2011, with it being quite low as compared to 1991. The contradiction of increase in overall sex ratio and decrease in child sex ratio in 2011 needs further study, as child sex ratio is often considered as a more reliable indicator of gender relations. The child-

Table 3.13 Some indicators of demographic profile of Madhya Pradesh during 1991

Sector	Average family size	Sex ratio	Child sex ratio
Total (R + U)	5.80	912.46	941.31
Rural	5.85	921.20	944.26
Urban	5.65	887.06	931.11

Source: Census of India 1991

Table 3.14 Some indicators of demographic profile of Madhya Pradesh in 2001

Sector	Average family zize	Sex ratio	Child sex ratio	Child-woman ratio	Dependency ratio	Young dependency ratio	Aged dependency ratio
Total	5.53	919.24	932.35	531	84.32	71.21	13.11
Rural	5.55	927.00	939.43	586	90.89	76.81	14.08
Urban	5.48	898.01	907.21	398	68.22	57.49	10.73

Source: Census of India 2001

Table 3.15 Some indicators of demographic profile of Madhya Pradesh in 2011

Sector	Average family size	Sex ratio	Child sex ratio	Child-woman ratio	Dependency ratio	Young dependency ratio	Aged dependency ratio
Total	4.81	930.93	917.86	227	71	57	13
Rural	4.74	935.86	922.80	243	77	63	14
Urban	5.00	918.15	901.47	185	56	45	12

Source: Census of India 2011

woman ratio in 2011 steeply declined to more than half vis-à-vis 2001. It became 227 in 2011 with it being 185 for urban areas. The dependency ratios of 2011 also decreased as compared to 2001. The dependency ratio and young dependency ratio during 2011 were 71 and 57, respectively, which were higher in rural areas and lower in urban areas.

3.3.2 *Morena District*

Morena district had alarmingly low rates of sex ratio and higher family sizes as compared to national average and state average in 1991 (as seen in Table 3.16). The average family size was 7.35 for overall Morena district, somewhat higher for rural areas (7.55) and slightly lower for urban areas (6.73). Sex ratio of Morena district was 808, with rural Morena being even lower (806). Child sex ratio was considerably higher (857) as compared to overall sex ratio.

The average family size in 2001 somewhat declined from 1991 levels to 6.7. Paradoxically, there were an increase in sex ratio and a decrease in child sex ratio (822 and 837, respectively) as compared to 1991 figures. Urban Morena district had

Table 3.16 Some indicators of demographic profile of Morena district in 1991

Sector	Average family size	Sex ratio	Child sex ratio
Total (R + U)	7.35	807.97	856.93
Rural	7.55	806.13	856.67
Urban	6.73	814.28	857.92

Source: Census of India 1991

Table 3.17 Some indicators of demographic profile of Morena district in 2001

Sector	Average family size	Sex ratio	Child sex ratio	Child-woman ratio	Dependency ratio	Young dependency ratio	Aged dependency ratio
Total (R + U)	6.70	822.14	837.46	601	92.21	79.69	12.52
Rural	6.74	816.89	839.16	640	95.90	82.69	13.22
Urban	6.56	841.52	830.25	479	79.85	69.67	10.19

Source: Census of India 2001

higher sex ratio (840 vis-à-vis average of 822) and lower child sex ratio as compared to rural Morena district. The child-woman ratio of Morena district in 2001 was quite high, 601, as compared to state average of 531.Urban parts of Morena district has relatively lower child-woman ratio (479) (Table 3.17). This district has extremely high dependency ratio, particularly in its rural areas (92 and 96 percentages, respectively) which are higher than the corresponding state shares. It also has high young dependency ratio (80). This again is higher for its rural areas and lower for its urban areas.

The decline in average size of family continues from 1991 to 2011. It becomes 5.42 for overall Morena district, which is high for its urban areas (6) as compared to its rural areas (5.26) (Table 3.18). The increase in overall sex ratio and decrease in child sex ratio continue (840 and 829, respectively). All sex ratios are quite low compared to either national or state averages of sex ratios. The child-woman ratio declined from 601 in 2001 to 452 in 2011. However, its figures are quite higher compared to the state figures. The dependency ratios declined during 2001–2011. The dependency ratio is 73 and young dependency ratio is 61. Urban areas of this district have lower dependency ratios as compared to its rural areas or overall figures.

3.3.3 *Tehsils of Morena District: Selected Demographic Indicators*

The average family size in different tehsils or sub-districts of Morena ranged from 7.68 to 6.91 in 1991 (Table 3.19). But by 2011, it had reduced to a range between 5.86 and 4.81 (Table 3.21). The sex ratio and child sex ratio have also shown significant variations across the administrative units. In all the census (1991, 2001 and

Table 3.18 Some indicators of demographic profile of Morena district, 2011

Sector	Average family size	Sex ratio	Child sex ratio	Child-woman ratio	Dependency ratio	Young dependency ratio	Aged dependency ratio
Total (R + U)	5.42	840.08	828.84	452	73	61	12
Rural	5.26	834.49	828.88	481	76	63	13
Urban	6.02	858.07	828.68	368	64	53	11

Source: Census of India 2011

Table 3.19 Selected demographic indicators in tehsils of Morena district, 1991

	Sector	Average family size	Sex ratio	Child sex ratio
Ambah	Total (R + U)	7.41	812.27	850.80
	Rural	7.51	812.20	855.51
	Urban	7.32	812.33	846.75
Porsa	Total (R + U)	7.15	830.28	855.89
	Rural	7.17	830.96	855.29
	Urban	7.14	829.69	856.41
Morena	Total (R + U)	7.53	784.40	842.75
	Rural	7.88	773.79	836.04
	Urban	7.35	790.31	846.65
Joura	Total (R + U)	7.68	797.06	864.52
	Rural	7.75	795.70	864.72
	Urban	7.63	798.25	864.34
Kailaras	Total (R + U)	7.68	821.83	865.61
	Rural	7.68	821.83	865.61
	Urban	7.68	821.83	865.61
Sabalgarh	Total (R + U)	6.91	835.32	874.12
	Rural	6.98	836.83	872.24
	Urban	6.85	834.15	875.60

Source: Census of India 1991

2011), rural Morena reported the lowest sex ratio, but it has increased over time from 774 in 1991 to 778 in 2001 to 806 in 2011. Rural Morena also had the lowest child sex ratio among all the administrative units of the district (Tables 3.19, 3.20 and 3.21).

3.4 Share of Social Groups, Morena

The social groupwise composition of a population is an important indicator of the social context of the area. More importantly, given the relationship between social group status and economic status of households in India, it presents an important

Table 3.20 Selected demographic indicators in tehsils of Morena district, 2001

	Sector	Average family size	Sex ratio	Child sex ratio
Ambah	Total (R + U)	6.67	828.36	819.28
	Rural	6.70	823.86	822.11
	Urban	6.53	851.47	802.47
Porsa	Total	6.58	849.89	851.95
	Rural	6.55	848.05	851.29
	Urban	6.74	859.20	855.73
Morena	Total (R + U)	6.77	805.51	810.46
	Rural	6.93	787.86	807.01
	Urban	6.51	836.78	817.54
Joura	Total (R + U)	6.80	808.32	849.94
	Rural	6.82	805.86	849.49
	Urban	6.69	832.36	854.74
Kailaras	Total	6.69	831.51	861.06
	Rural	6.74	830.47	857.83
	Urban	6.43	837.80	884.22
Sabalgarh	Total (R + U)	6.47	848.67	869.74
	Rural	6.44	848.76	877.67
	Urban	6.58	848.34	839.35

Source: Census of India 2001

Table 3.21 Selected demographic indicators in tehsils of Morena district, 2011

	Sector	Average family size	Sex ratio	Child sex ratio
Ambah	Total (R + U)	5.37	856.06	827.02
	Rural	5.22	850.79	830.51
	Urban	6.14	880.09	808.70
Porsa	Total (R + U)	5.66	861.83	842.81
	Rural	5.62	855.42	833.14
	Urban	5.84	893.42	894.86
Morena	Total (R + U)	5.86	823.59	806.42
	Rural	5.73	806.13	800.32
	Urban	6.05	847.80	816.17
Joura	Total (R + U)	5.20	831.25	839.34
	Rural	5.11	828.43	838.02
	Urban	6.28	858.35	852.90
Kailaras	Total (R + U)	5.22	848.85	843.85
	Rural	5.12	847.13	844.60
	Urban	5.90	859.93	838.50
Sabalgarh	Total (R + U)	4.81	856.44	845.13
	Rural	4.57	854.32	847.87
	Urban	5.82	863.65	834.78

Source: Census of India 2011

Table 3.22 Share of SC and ST population in blocks of Morena district

	Percentage of SC		Percentage of ST	
Block name	2001	2011	2001	2011
Ambah	24.07	25.63	0.57	0.08
Porsa	23.51	24	0.14	0.04
Morena	21.62	22.52	0.3	0.15
Pahadgarh	18.38	17.14	2.92	4.02
Joura	18.61	18.7	0.34	0.52
Kailaras	18.82	19.44	1.94	1.77
Sabalgarh	21.99	23.45	3.86	1.68

Source: Census of India 2001 2011

aspect of the socio-economic scenario of the region. For determining social composition of the population, the respective shares of scheduled castes (SC), scheduled tribes (ST) and others (both OBCs and others) are computed for all, males and females and across rural, urban and total population totals of the respective spatial units of analysis. This is done across three time periods – 1991, 2001 and 2011. The data presented in Table 3.22 suggest that among the tehsils, share of SC population is high in Ambah, Porsa and Sabalgarh, whereas the share of ST population is higher in Pahadgarh and Sabalgarh. The fluctuating share of the ST population in these two blocks shows significant changes in the social composition of the population.

3.4.1 *Madhya Pradesh*

In 1991, broadly 65% of the total population of Madhya Pradesh comprises of socially higher castes, i.e. others. So, others comprise the largest social group in Madhya Pradesh in 1991, followed by ST and SC population (around 20 and 15%, respectively). Urban Madhya Pradesh however varies, as around 81% of the population (roughly) are others, with only few percentage of SC and ST population in urban areas (around 14 and 4.2 percentages, respectively). Rural Madhya Pradesh on the contrary has sizeable share of ST population of 25%, though others still comprise the largest social group (around 58%), as could be seen in Table 3.23.

In 2001, the shares of different social groups in Madhya Pradesh follow similar trends as of the 1991 figures without any major deviations (Table 3.24). The social composition and ranking of population in Madhya Pradesh in 2011 remains broadly the same as of 1991 and 2001 trends. However a minute increase in ST population was observed from 2001 to 2011 for all, rural and urban (Table 3.25).

Table 3.23 Share of different social groups to total population for Madhya Pradesh, 1991

Social groups	SC			ST			Others		
Total	15.40	15.54	15.24	19.94	19.34	20.58	64.67	65.11	64.17
Rural	15.81	16.02	15.59	25.25	24.56	26.01	58.94	59.43	58.40
Urban	14.17	14.16	14.19	4.22	4.20	4.23	81.61	81.64	81.58

Source: Census of India 1991

Table 3.24 Share of different social groups to total population in Madhya Pradesh, 2001

Social groups	SC			ST			Others		
Sector	All	Male	Female	All	Male	Female	All	Male	Female
Total	15.17	15.28	15.05	20.27	19.70	20.89	64.56	65.02	64.06
Rural	15.58	15.76	15.39	25.79	25.11	26.52	58.63	59.13	58.09
Urban	14.03	13.97	14.10	4.93	4.89	4.97	81.04	81.14	80.93

Source: Census of India 2001

Table 3.25 Share of different social groups to total population in Madhya Pradesh, 2011

Social groups	SC			ST			Others		
Sector	All	Male	Female	All	Male	Female	All	Male	Female
Total	15.62	15.71	15.52	21.09	20.52	21.70	63.29	63.77	62.78
Rural	15.73	15.88	15.57	27.16	26.47	27.90	57.10	57.64	56.53
Urban	15.32	15.26	15.38	5.18	5.08	5.29	79.50	79.65	79.33

Source: Census of India 2011

3.4.2 Morena District

In 1991, roughly three-fourths of the population of Morena district constitutes others (78.4%). This district almost lacks any ST population vis-à-vis state average of 25% ST population. Rather SC is the second most dominant social group (around 21%), having a slightly higher than average state share (see Table 3.26). There are negligible sector-wise differences in social composition. The trends of social composition of Morena district in 2001 are similar to that of 1991, with very miniscule increase in SC and ST percentage in 2001 vis-à-vis 1991 (Table 3.27). In 2011, though the trend remains similar to 1991 and 2001, the slow increase of SC and ST population continues, with corresponding decline of others (Table 3.28).

3.4.3 Tehsils of Morena District

In 1991, most of the tehsils of Morena district have largest share of population of others (around 75%). The next largest social group is SC (around 21–23%), and ST population has negligible share. Porsa, Morena, Joura and Kailaras tehsils have

Table 3.26 Share of different social groups to total population in Morena district, 1991

Social groups	SC			ST			Others		
Total	20.98	21.07	20.88	0.67	0.64	0.71	78.35	78.29	78.41
Rural	20.92	20.98	20.84	0.79	0.75	0.84	78.29	78.27	78.32
Urban	21.21	21.36	21.02	0.25	0.25	0.25	78.54	78.39	78.72

Source: Census of India 1991

Table 3.27 Share of different social groups to total population in Morena district, 2001

Social groups	SC			ST			Others		
Sector	All	Male	Female	All	Male	Female	All	Male	Female
Total	21.08	21.14	21.01	0.81	0.78	0.85	78.11	78.08	78.14
Rural	21.29	21.38	21.19	0.93	0.88	0.98	77.78	77.74	77.83
Urban	20.30	20.25	20.36	0.40	0.41	0.39	79.30	79.34	79.25

Source: Census of India 2001

Table 3.28 Share of different social groups to total population in Morena district, 2011

Social groups	SC			ST			Others		
Sector	All	Male	Female	All	Male	Female	All	Male	Female
Total	21.44	21.39	21.50	0.87	0.84	0.90	77.69	77.77	77.60
Rural	21.50	21.45	21.56	0.96	0.92	1.01	77.54	77.63	77.43
Urban	21.25	21.21	21.31	0.57	0.56	0.57	78.18	78.23	78.12

Source: Census of India 2011

even higher shares of others (ranging from 78% to 80%) (Table 3.29). The social composition in 2001 remained broadly similar to that of 1991 (Table 3.30). While in 2011, existing trends of 2001 persisted, but a minor increase in share of SC population was observed in few of the tehsils (as seen in Table 3.31).

3.5 Share of Different Religious Groups

The religious composition is computed by percentage shares of Hindus, Muslims, Christians, Sikhs, Buddhists, Jains, other religions (or religion not stated) and no religion groups to total population for total, rural and urban for different spatial units and across three different time periods – 1991, 2001 and 2011.

3.5.1 Madhya Pradesh

In 1991, Hindus comprised the largest religious group in Madhya Pradesh, followed by Muslims (93 and 5 percentages, respectively). There are negligible shares of population by other religions. Rural Madhya Pradesh has even higher percentage of

Table 3.29 Share of different social groups to total population in tehsils of Morena district, 1991

Tehsil name	Sector	SC			ST			Others		
Ambah	Total	23.40	23.43	23.36	0.05	0.04	0.05	76.55	76.52	76.59
	Rural	23.97	24.02	23.92	0.02	0.02	0.02	76.00	75.96	76.05
	Urban	22.91	22.94	22.88	0.07	0.07	0.08	77.02	77.00	77.04
Porsa	Total	21.63	21.73	21.50	0.01	0.02	0.01	78.36	78.25	78.49
	Rural	21.81	21.90	21.70	0.01	0.02	0.00	78.18	78.08	78.30
	Urban	21.47	21.59	21.33	0.02	0.02	0.01	78.51	78.39	78.66
Morena	Total	21.24	21.19	21.30	0.29	0.31	0.26	78.47	78.50	78.44
	Rural	20.11	20.01	20.25	0.29	0.33	0.25	79.59	79.67	79.50
	Urban	21.86	21.85	21.87	0.29	0.30	0.27	77.85	77.85	77.86
Joura	Total	18.64	18.89	18.33	1.30	1.22	1.40	80.06	79.89	80.27
	Rural	18.78	19.02	18.47	1.38	1.29	1.49	79.85	79.69	80.04
	Urban	18.52	18.78	18.21	1.23	1.16	1.33	80.24	80.07	80.46
Kailaras	Total	19.09	19.15	19.02	1.71	1.57	1.88	79.20	79.29	79.10
	Rural	19.09	19.15	19.02	1.71	1.57	1.88	79.20	79.29	79.10
	Urban	19.09	19.15	19.02	1.71	1.57	1.88	79.20	79.29	79.10
Sabalgarh	Total	22.97	23.03	22.91	1.35	1.28	1.42	75.68	75.69	75.67
	Rural	23.81	23.88	23.73	1.51	1.44	1.61	74.67	74.68	74.66
	Urban	22.32	22.37	22.27	1.21	1.17	1.27	76.46	76.46	76.46

Source: Census of India 1991

Table 3.30 Share of different social groups to total population in tehsils of Morena district, 2001

	Social groups	SC			ST			Others		
Tehsil name	Sector	All	Male	Female	All	Male	Female	All	Male	Female
Ambah	Total	23.18	23.20	23.14	0.52	0.53	0.50	76.31	76.27	76.35
	Rural	24.08	24.12	24.04	0.58	0.59	0.57	75.33	75.29	75.39
	Urban	18.58	18.48	18.70	0.19	0.20	0.19	81.23	81.33	81.12
Porsa	Total	22.67	22.72	22.62	0.15	0.14	0.15	77.18	77.14	77.23
	Rural	23.22	23.28	23.16	0.15	0.15	0.16	76.62	76.57	76.69
	Urban	19.91	19.87	19.95	0.10	0.10	0.10	80.00	80.03	79.95
Morena	Total	22.14	22.04	22.26	0.35	0.36	0.34	77.51	77.60	77.39
	Rural	22.03	21.96	22.12	0.31	0.31	0.32	77.66	77.73	77.57
	Urban	22.32	22.17	22.51	0.43	0.46	0.39	77.25	77.38	77.10
Joura	Total	18.41	18.60	18.17	1.36	1.28	1.47	80.23	80.12	80.36
	Rural	18.43	18.62	18.19	1.42	1.33	1.54	80.15	80.05	80.27
	Urban	18.23	18.42	18.00	0.78	0.74	0.83	80.99	80.84	81.17
Kailaras	Total	18.76	18.93	18.56	1.78	1.69	1.89	79.46	79.38	79.55
	Rural	18.80	18.98	18.58	2.01	1.91	2.14	79.19	79.11	79.28
	Urban	18.54	18.59	18.47	0.38	0.37	0.39	81.09	81.04	81.14
Sabalgarh	Total	21.18	21.39	20.93	1.23	1.17	1.31	77.59	77.44	77.76
	Rural	22.68	22.92	22.40	1.48	1.40	1.57	75.84	75.69	76.03
	Urban	16.21	16.32	16.07	0.43	0.43	0.42	83.36	83.25	83.50

Source: Census of India 2001

Table 3.31 Share of different social groups to total population in tehsils of Morena district, 2011

	Social groups	SC			ST			Others		
Tehsil name	Sector	All	Male	Female	All	Male	Female	All	Male	Female
Ambah	Total	24.49	24.42	24.58	0.13	0.12	0.13	75.38	75.46	75.29
	Rural	25.63	25.57	25.69	0.08	0.08	0.09	74.29	74.35	74.22
	Urban	19.43	19.16	19.72	0.31	0.33	0.29	80.26	80.50	79.99
Porsa	Total	23.20	23.15	23.27	0.07	0.07	0.08	76.72	76.78	76.66
	Rural	24.00	23.91	24.11	0.04	0.05	0.04	75.95	76.04	75.85
	Urban	19.34	19.40	19.28	0.22	0.20	0.24	80.44	80.41	80.48
Morena	Total	22.33	22.12	22.57	0.25	0.24	0.25	77.43	77.64	77.17
	Rural	21.59	21.33	21.91	0.13	0.13	0.14	78.28	78.54	77.95
	Urban	23.32	23.22	23.45	0.40	0.40	0.41	76.27	76.38	76.15
Joura	Total	18.22	18.31	18.11	1.90	1.84	1.98	79.87	79.85	79.91
	Rural	18.06	18.15	17.94	1.97	1.90	2.05	79.98	79.95	80.00
	Urban	19.78	19.88	19.66	1.30	1.30	1.30	78.92	78.81	79.04
Kailaras	Total	19.13	19.18	19.07	1.67	1.61	1.73	79.20	79.21	79.20
	Rural	19.44	19.52	19.34	1.77	1.72	1.84	78.79	78.77	78.82
	Urban	17.18	17.04	17.35	0.99	0.93	1.07	81.82	82.03	81.58
Sabalgarh	Total	21.94	22.06	21.80	1.55	1.48	1.64	76.50	76.45	76.56
	Rural	23.45	23.55	23.34	1.68	1.59	1.79	74.87	74.86	74.88
	Urban	16.87	17.04	16.67	1.12	1.13	1.12	82.01	81.83	82.21

Source: Census of India 2011

Table 3.32 Shares of different religious groups to total population in Madhya Pradesh, 1991 *(in percentages)*

Sector	Hindus	Muslims	Christians	Sikhs	Buddhists	Jains	Other	No Religion
Total	92.80	4.96	0.64	0.24	0.33	0.74	0.09	0.19
Rural	96.24	2.33	0.53	0.08	0.27	0.25	0.12	0.18
Urban	81.36	13.67	1.04	0.77	0.52	2.36	0.03	0.25

Source: Census of India 1991

Hindus (96%) (see Table 3.32.). However Urban Madhya Pradesh had somewhat lower share of Hindus and prominent share of Muslims (81 and 14 percentages, respectively). The religious composition of population of Madhya Pradesh in 2001 remained same as that in 1991; however a slight increase in Muslim share was observed in 2001 and a decline in Hindus share (Table 3.33). The trends in 2011 remained broadly the same (Table 3.34).

3.5.2 *Morena District*

The religious composition of Morena district is almost similar to Madhya Pradesh. Hindus comprise the largest religious group followed by Muslims (95 and 4 percentages, respectively). Like urban M.P., urban Morena district has somewhat

Table 3.33 Shares of different religious groups to total population in Madhya Pradesh, 2001 *(in percentages)*

Sector	Hindus	Muslims	Christians	Sikhs	Buddhists	Jains	Other
Total	91.15	6.37	0.28	0.25	0.35	0.90	0.68
Rural	95.07	3.16	0.13	0.13	0.30	0.28	0.91
Urban	80.25	15.28	0.70	0.59	0.49	2.63	0.03

Source: Census of India 2001

Table 3.34 Shares of different religious groups to total population in Madhya Pradesh, 2011 *(in percentages)*

Sector	Hindus	Muslims	Christians	Sikhs	Buddhists	Jains	Other	No Religion
Total	90.89	6.57	0.29	0.21	0.30	0.78	0.83	0.13
Rural	94.80	3.20	0.17	0.10	0.25	0.21	1.13	0.13
Urban	80.62	15.40	0.62	0.48	0.42	2.28	0.04	0.15

Source: Census of India 2011

Table 3.35 Shares of different religious groups to total population in Morena district, 1991 *(in percentages)*

Sector	Hindus	Muslims	Christians	Sikhs	Buddhists	Jains	Other	Not stated
Total	94.89	4.13	0.01	0.28	0.07	0.40	0.01	0.22
Rural	96.82	2.47	0.00	0.33	0.09	0.06	0.01	0.23
Urban	87.43	10.56	0.03	0.09	0.01	1.72	0.00	0.16

Source: Census of India 1991

higher share of Muslims and a slightly lower share of Hindus (10 and 87 percentages, respectively) (Table 3.35). Trends in 2001 remained broadly the same (Table 3.36). While trends in 2011 remained the same, a slight decline in Muslim population, especially in urban Morena district, was observed (Table 3.37).

3.5.3 *Tehsils of Morena District*[2]

Hindus comprise the largest religious group across all tehsils of Morena district in 2011, ranging from 93% to 97%. Muslims rank next in terms of population shares, and other religions are non-existent. However rural and urban parts of tehsils have slight deviations, with the urban parts having somewhat higher shares of Muslim population. This is specially for the tehsils of Joura, Porsa and Sabalgarh. Urban parts of the tehsils of Ambah and Porsa have around 2–3% Jain population (Table 3.38).

[2] Census data on population by different religious groups are available at tehsil level only in 2011.

Table 3.36 Shares of different religious groups to total population in Morena district, 2001 *(in percentages)*

Sector	Hindus	Muslims	Christians	Sikhs	Buddhists	Jains	Other
Total	95.65	3.76	0.03	0.07	0.02	0.45	0.00
Rural	97.35	2.48	0.02	0.05	0.02	0.06	0.00
Urban	89.46	8.43	0.07	0.10	0.02	1.89	0.01

Source: Census of India 2001

Table 3.37 Shares of different religious groups to total population in Morena district, 2011 *(in percentages)*

Sector	Hindus	Muslims	Christians	Sikhs	Buddhists	Jains	Other	Not stated
Total	95.40	3.87	0.06	0.03	0.16	0.34	0.00	0.14
Rural	97.02	2.55	0.04	0.02	0.19	0.03	0.00	0.14
Urban	90.24	8.07	0.11	0.05	0.08	1.32	0.01	0.12

Source: Census of India 2011

Table 3.38 Percentages of different religious groups to total population in tehsils of Morena district, 2011 *(in percentages)*

Tehsil		Hindus	Muslims	Christians	Sikhs	Buddhists	Jains	Other	Not stated
Ambah	Total	97.33	1.83	0.05	0.01	0.01	0.66	0.01	0.11
	Rural	98.78	1.02	0.04	0.01	0.00	0.04	0.01	0.10
	Urban	90.83	5.45	0.07	0.01	0.04	3.41	0.00	0.19
Porsa	Total	97.22	2.24	0.06	0.01	0.01	0.38	0.01	0.07
	Rural	98.85	1.00	0.05	0.01	0.01	0.00	0.00	0.07
	Urban	89.36	8.24	0.12	0.03	0.00	2.19	0.01	0.06
Morena	Total	93.77	5.25	0.07	0.07	0.13	0.47	0.01	0.24
	Rural	96.20	3.16	0.04	0.07	0.19	0.02	0.01	0.33
	Urban	90.47	8.07	0.12	0.07	0.06	1.08	0.01	0.13
Joura	Total	94.69	4.74	0.03	0.01	0.18	0.26	0.01	0.09
	Rural	95.54	4.08	0.03	0.01	0.17	0.08	0.00	0.08
	Urban	86.61	10.92	0.07	0.03	0.27	1.96	0.02	0.13
Kailaras	Total	96.83	3.02	0.04	0.01	0.02	0.02	0.00	0.06
	Rural	97.60	2.30	0.03	0.01	0.01	0.00	0.00	0.05
	Urban	91.88	7.65	0.11	0.01	0.07	0.15	0.00	0.13
Sabalgarh	Total	96.02	3.11	0.06	0.02	0.69	0.01	0.00	0.09
	Rural	97.39	1.60	0.04	0.01	0.86	0.00	0.00	0.09
	Urban	91.38	8.20	0.12	0.07	0.14	0.04	0.00	0.06

Source: Census of India 2011

3.6 Share of Workers by Industries/Occupation

The Census of India classifies both main and marginal workers[3] on the basis of their industrial classification into four categories – cultivator, agricultural labourer, household industry worker and other workers. The share of workers by the four major industrial groups out of the total workers (i.e. both main and marginal workers) is computed for all, males and females and across rural, urban and total population totals of the respective spatial units of analysis. This is done across two time periods – 2001 and 2011.

3.6.1 *Madhya Pradesh*

Out of the total workers in M.P. in 2001, around 42% are cultivators, 14% agricultural labourers, only 4% household industry workers and 25% other workers. Rural and urban scenario of industrial composition of workforce varies. In rural areas, nearly half of all workers work as cultivators, followed by around one sixth share of agricultural workers and only 10% of other workers. After cultivators, nearly one third rural men work as agricultural labourers. In urban areas of M.P., workers are mostly engaged as other workers (82%) (Table 3.39).

Cultivators and agricultural labourers individually constitute about one third shares of workers in 2011, followed by other workers (around 24%).In rural areas, major share of women work either as cultivator or agricultural labourer. Most of the workers in urban areas are engaged as other workers (75%) (Table 3.40).

Table 3.39 Percentage shares of total workers (main + marginal) by major industrial groups in Madhya Pradesh, for all, male and female – 2001

	Cultivator			Agricultural labourer			Household ind. worker			Other worker		
Sector	All	Male	Female	All	Male	Female	All	Male	Female	All	Male	Female
Total	42.79	42.50	43.29	13.64	23.97	10.76	4.01	3.21	5.36	24.51	32.57	10.91
Rural	51.36	54.56	46.86	16.16	30.79	8.19	3.41	2.85	4.19	11.09	14.91	5.73
Urban	6.22	5.60	8.96	2.86	3.12	35.48	6.56	4.29	16.57	81.82	86.60	60.73

Source: Census of India 2001

[3]Main workers are workers 'who had worked for the major part of the reference period i.e. 6 months or more', while marginal workers are those workers 'who had not worked for the major part of the reference period i.e. less than 6 months'.

Table 3.40 Percentage shares of total workers (main + marginal) by major industrial groups in Madhya Pradesh, for all, male and female – 2011

	Cultivator			Agricultural labourer			Household ind. worker			Other worker		
Sector	All	Male	Female	All	Male	Female	All	Male	Female	All	Male	Female
Total	31.18	32.71	28.47	38.61	31.32	51.47	2.82	2.41	3.54	23.67	30.02	12.48
Rural	38.33	42.74	31.82	47.30	40.77	56.97	2.18	1.80	2.74	9.57	12.24	5.61
Urban	5.40	5.38	5.50	7.31	5.57	13.76	5.13	4.07	9.05	74.51	78.52	59.62

Source: Census of India 2011

Table 3.41 Percentage shares of total workers (main + marginal) by major industrial groups in Morena, for all, male and female – 2001

	Cultivator			Agricultural labourer			Household ind. worker			Other worker		
Sector	All	Male	Female	All	Male	Female	All	Male	Female	All	Male	Female
Total	56.80	61.17	46.05	6.86	6.50	6.82	1.97	1.43	3.30	29.75	27.76	34.66
Rural	65.20	73.14	48.29	7.61	7.78	5.27	1.68	1.17	2.78	20.21	14.51	32.35
Urban	10.98	10.53	14.23	2.76	1.05	28.82	3.53	2.52	10.75	81.81	83.81	67.51

Source: Census of India 2001

Table 3.42 Percentage shares of total workers (main + marginal) by major industrial groups in Morena, for all, male and female – 2011

	Cultivator			Agricultural labourer			Household ind. worker			Other worker		
Sector	All	Male	Female	All	Male	Female	All	Male	Female	All	Male	Female
Total	45.07	49.70	28.93	21.93	20.86	25.63	2.16	1.54	4.30	23.48	23.85	22.21
Rural	52.89	60.21	31.18	25.52	24.86	27.48	1.92	1.16	4.17	12.79	10.89	18.42
Urban	11.96	12.77	4.55	6.70	6.82	5.55	3.18	2.90	5.74	68.77	69.38	63.22

Source: Census of India 2011

3.6.2 *Morena District*

Cultivators form the largest share of workers in Morena district in 2001, followed by other workers (57 and 30%, respectively). In rural areas, cultivators comprise even larger share of workers for both male and female workers. Contrarily in urban areas of Morena district, only 11% work as cultivators, and workers are mostly engaged as other workers (82%) (Table 3.41). Trends of workforce in Morena district similar to 2001 are observed in 2011 (Table 3.42). The sharp rural-urban differences in the industrial distribution of workers are an important aspect of the livelihoods scenario in Morena district.

3.6.3 Tehsils of Morena District

In almost all the tehsils of Morena district, cultivators and other workers are the two major industries employing workers. However, Ambah, Porsa and Morena have around 45% workers engaged and cultivators and around 40% as other workers. Joura, Sabalgarh and Kailaras tehsils however have higher share of cultivators, both for overall workers and for its rural areas (roughly 70%). In almost all the tehsils, the share of agricultural workers in rural areas was around 8% or less, showing a relatively lesser dependence on wage labour market than in the state. While other workers comprise largest share of workers in urban areas of all tehsils, major share of women urban workers are also employed as agricultural labourers. This is specially for the tehsils of Kailaras, Joura and Morena (ranging from 34% to 44%) (Table 3.43).

Broad trends of workforce shared by different industrial groups in 2011 remain same as that of 2001. Cultivators and other workers comprise the largest share of workers for both males and females and both sectors. However substantial share of workers in 2011 are also engaged as agricultural labourers (roughly 20%) (as seen in Table 3.44).

3.7 Conclusion

This chapter provides a broad overview of the socio-economic scenario of the study area. In order to understand the socio-economic condition of the study area, a range of social and economic indicators were selected and discussed in relation to the state as well as the district averages. The changes in the socio-economic conditions were investigated through the inter-temporal changes in the selected indicators, although the changes of administrative boundaries pose some problems for the inter-temporal analysis. Further, wherever possible a disaggregated picture was presented with data from the block or tehsil level. Given the ecological conditions of the region, tehsil-level picture is of great importance. The various indicators presented in this chapter bring out the social and educational backwardness of the region. Not only that literacy rates are low, there is a substantial gap in the male and female literacy rates. Sex ratio and child sex ratio point to significant female disadvantage in the region. There is a significant share of the marginalised social groups although religious groupwise diversity is found to be low. Majority of the rural workers were found to be dependent in agriculture as cultivators. Between 2001 and 2011, there has been a substantial increase in the share of agricultural labour in the rural areas of Morena district.

The analysis presented in this chapter has implications for the overall understanding of the relationship between land degradation and socio-economic development. To understand the relationship, it is important to contextualise the levels of economic development in the disaggregated units (such as tehsils) with the overall

Table 3.43 Percentage shares of workers (main + marginal) by major industrial groups in tehsils of Morena, for all, male and female – 2001

		Cultivator			Agricultural labourer			Household ind. worker			Other worker		
Tehsil name	Sector	All	Male	Female	All	Male	Female	All	Male	Female	All	Male	Female
Ambah	Total	49.93	63.80	14.70	8.26	3.72	5.87	1.66	0.92	3.53	37.48	23.76	72.33
	Rural	54.90	72.02	15.35	8.95	4.25	4.59	1.39	0.58	3.25	31.80	14.58	71.58
	Urban	12.95	14.37	3.91	3.16	0.49	27.34	3.69	3.01	8.05	79.77	78.97	84.92
Porsa	Total	46.42	63.41	9.19	7.56	3.60	5.55	1.74	1.06	3.23	41.81	24.52	79.69
	Rural	51.08	71.75	9.77	7.93	4.11	4.49	1.50	0.78	2.94	36.75	15.57	79.07
	Urban	15.75	18.73	2.24	5.10	0.86	18.36	3.32	2.57	6.72	75.13	72.47	87.16
Morena	Total	44.38	46.43	37.53	6.80	5.76	10.73	2.47	1.69	5.09	41.91	43.05	38.13
	Rural	59.47	66.15	42.24	8.89	8.17	6.85	1.91	1.28	3.55	23.84	20.24	33.14
	Urban	5.01	5.11	4.16	1.36	0.73	38.23	3.93	2.54	16.03	89.04	90.83	73.45
Joura	Total	66.67	70.28	57.86	7.93	8.92	8.46	2.46	1.93	3.76	16.62	16.61	16.65
	Rural	70.69	75.88	58.85	7.91	9.60	7.81	2.38	1.83	3.63	12.34	10.92	15.59
	Urban	11.64	10.80	18.76	8.21	1.64	34.05	3.57	2.97	8.66	75.11	77.05	58.61
Kailaras	Total	73.63	70.67	78.73	4.88	8.40	3.67	1.35	1.10	1.77	14.83	20.51	5.04
	Rural	79.48	79.04	80.17	5.14	9.38	2.74	1.08	0.93	1.30	8.61	11.55	4.06
	Urban	11.98	11.50	16.58	2.12	1.48	44.03	4.22	2.31	22.28	80.34	83.85	47.24
Sabalgarh	Total	69.45	67.02	74.49	4.34	8.38	3.16	1.03	1.03	1.03	19.52	25.50	7.15
	Rural	78.65	79.83	76.55	4.89	10.12	2.46	0.89	0.84	0.97	9.10	11.67	4.57
	Urban	26.59	21.47	53.05	1.77	2.20	10.42	1.69	1.71	1.59	68.10	74.71	33.97

Source: Census of India 2001

Table 3.44 Percentage shares of workers (main + marginal) by major industrial groups in tehsils of Morena, for all, male and female – 2011

		Cultivator			Agricultural labourer			Household ind. worker			Other worker		
Tehsil name	Sector	All	Male	Female	All	Male	Female	All	Male	Female	All	Male	Female
Ambah	Total	36.51	51.40	5.94	20.37	24.84	11.20	2.51	1.26	5.07	23.25	17.43	35.22
	Rural	39.34	57.70	6.09	21.73	27.37	11.52	2.36	0.92	4.97	18.49	9.58	34.63
	Urban	16.44	18.40	2.88	10.74	11.60	4.77	3.58	3.08	7.04	57.10	58.56	46.94
Porsa	Total	46.47	59.01	9.12	18.75	20.88	12.42	1.61	0.79	4.04	20.42	15.74	34.36
	Rural	51.02	66.17	9.67	20.69	23.38	13.33	1.56	0.63	4.11	13.94	7.19	32.38
	Urban	18.25	20.89	3.03	6.78	7.57	2.20	1.87	1.63	3.26	60.60	61.29	56.62
Morena	Total	35.16	37.17	24.00	17.92	15.85	29.41	3.08	2.35	7.14	37.00	38.50	28.69
	Rural	51.30	56.08	30.36	25.71	23.01	37.55	2.67	1.58	7.44	15.85	15.80	16.06
	Urban	8.52	9.00	4.05	5.06	5.19	3.87	3.76	3.49	6.22	71.92	72.31	68.30
Joura	Total	51.13	54.39	37.86	29.71	26.97	40.85	2.25	1.64	4.73	14.21	15.05	10.83
	Rural	54.42	58.56	38.73	30.98	28.20	41.52	2.17	1.53	4.59	9.70	9.75	9.53
	Urban	12.06	12.65	3.16	14.61	14.64	14.17	3.24	2.78	10.32	67.79	68.12	62.86
Kailaras	Total	56.94	57.04	56.43	22.21	20.85	29.12	0.86	0.73	1.52	17.28	18.60	10.62
	Rural	62.94	63.87	58.68	24.50	23.30	30.01	0.82	0.70	1.39	9.34	9.66	7.89
	Urban	10.88	11.34	3.41	4.67	4.46	8.05	1.17	0.95	4.63	78.20	78.41	74.88
Sabalgarh	Total	57.15	56.37	58.88	21.89	17.36	31.98	0.97	0.97	0.96	17.30	22.35	6.03
	Rural	64.14	65.81	61.07	25.01	20.77	32.79	0.80	0.75	0.90	7.89	10.33	3.41
	Urban	20.88	21.63	13.00	5.72	4.83	14.98	1.83	1.78	2.30	66.06	66.54	61.07

Source: Census of India 2011

patterns at the district and the state levels. Further, inter-tehsil differences (at sub-district levels) point to the specific patterns of socio-economic development in a ravine-affected region. Another insight from the exercise is to understand the overall nature of social differentiations in the extent of development, captured through a limited but significant set of variables. These insights have been used in interpreting the data generated through primary survey.

References

Census of India (1991) Final Population Totals. Available at https://data.gov.in/keywords/census-1991

Census of India (2001) General Population Tables. Available at https://censusindia.gov.in/Census_Data_2001/Census_data_finder/Census_Data_Finder.aspx

Census of India (2011) Census of India 2011, series – 24 Part Xii-A, village and town directory, 2011, directorate of census operations, Madhya Pradesh, Morena (Part A and B). Available at http://censusindia.gov.in/2011census/dchb/DCHB_A/23/2302_PART_A_DCHB_MORENA.pdfandhttp://censusindia.gov.in/2011census/dchb/2302_PART_B_DCHB_MORENA.pdf

Sen A (1999) Development as freedom. Oxford University Press, Delhi

Chapter 4
Land Degradation and Socio-economic Development: Linkages

Abstract Largely based on secondary data sources, a comparative analysis of the socio-economic scenario of the degraded and non-degraded villages in the district has been placed in the fourth chapter. The social, economic and educational aspects of development are mapped, and an attempt has been made to establish a link between the land degradation and rural development. Two key insights that have been brought out from the analysis of the distribution of villages across size classes of different attributes are as follows. Firstly, as in the case of other fragile ecological regions, spatial distribution of the population across the villages is an important aspect of the pattern of socio-economic development in this region. Secondly, the spatial pattern also gets reinforced by the linkages among the distribution of villages according to various indicators.

Keywords Socio-economic development · Linkages · Environmental resources · Livelihoods · Common property resources · Village shifting

4.1 Introduction

Economic development is the final outcome of a number of different activities. The development of a region is at times based on environmental resources, but the relationship is far from being straightforward. The availability of natural resources acts as an advantage for a region[1] (Gallup et al. 1999). There are many instances where the long-term development of an economy starts with economic exploitation of natural resources. Minerals, water, soil and forest wealth have acted as the drivers of growth in many countries. However, there are many instances where countries and

[1] Gallup et al.. 1999. Geography and economic development. *International regional science review*, *22*(2), pp.179–232. Gallup et al. (1999) conclude that 'Tropical regions are hindered in development relative to temperate regions, probably because of higher disease burdens and limitations on agricultural productivity' and 'Coastal regions, and regions linked to coasts by ocean-navigable waterways, are strongly favoured in development relative to the hinterlands'.

P. Pani, *Land Degradation and Socio-Economic Development*, Advances in Asian Human-Environmental Research,
https://doi.org/10.1007/978-3-030-42074-1_4

regions endowed with rich natural resources suffer from lack of development, a phenomenon termed as 'resource curse'. Difficult trains, frequent natural disasters or difficult climatic conditions also create significant barriers for economic progress.

Another aspect of the relationship between environment and development nexus is the impact of economic development on natural environment. Often the price of economic progress is destruction of the natural environment. This problem has many different dimensions – some of which originate from the 'externality problem'[2]. The true value of the environment cannot often be included in the economy, and, also, there is a divergence between collective welfare of the humankind and individual profit of those who control the resources. This leads to overexploitation of natural resources. The Environmental Kuznets Curve (EKC) argument suggests that the relationship between environmental degradation and per capita income is that of an inverse U-shaped curve. With initial increase in the per capita income, environmental degradation or pollution levels tend to increase; however, after a threshold level of per capita income is reached, countries can afford to adopt costlier but less degrading technologies, and thus environmental degradation gradually decreases[3].

The problem of environmental degradation in the developing economies is all the more complex, because poverty and lack of development restrict the choice of people and force them to adopt practices that are less sustainable. The short-term urgency of survival needs often takes precedence over the long-term objectives of sustainability. However, there is a counter view that poor people, because of their greater dependence on natural resources for survival, tend to value natural resources more than others[4] (Jodha 1998).

In the specific context of land degradation, the relationship is more complex. Often land degradation is the cause of poverty and backwardness. Again, poverty accentuates land degradation[5]. Poor peasants, in response to declining land

---

[2] Environmental externalities refer to 'the economic concept of uncompensated environmental effects of production and consumption that affect consumer utility and enterprise cost outside the market mechanism.

As a consequence of negative externalities, private costs of production tend to be lower than its "social" cost. It is the aim of the "polluter/user-pays" principle to prompt households and enterprises to internalise externalities in their plans and budgets' (OECD: Glossary of Statistical terms. Available at https://stats.oecd.org/glossary/detail.asp?ID=824).

[3] The EKC is also related to the structural transformation of the economy. Initially, economic progress is driven by industrialization, which tends to be more polluting than the later stage when less-polluting economic activities such as services predominate.

[4] For details see, Jodha NS. Common property resources and rural poor in dry regions of India. *Economic and political weekly*. 1986 Jul 5:1169–81.

[5] Explaining the 'cumulative causation' in the African context, a study argues: 'Many poor African pastoralists and farming households respond to declining land productivity by abandoning their existing degraded pasture and cropland, and moving to new lands for grazing and cultivation. Even if rural households choose to stay on degraded land, its declining productivity will be unable to support growing rural populations. Thus, some households will be forced to abandon existing agricultural areas in search of new land. However, without additional investments in soil conservation, this process will repeat itself. Eventually, overgrazing and cultivation will lead to land degradation, and the search for new pasture and cropland will begin again' (Barbier 2000: 351). For

productivity, exploit marginal land or use the same methods of cultivation that adversely affect soil health.

> As low potential lands are considered to be prone to chronic land degradation, then clearly the problems of resource management by poor rural households and human-induced soil degradation are linked in developing countries. Moreover, given that many marginal and resource-poor lands are also likely to have been previously forested lands, then a strong rural poverty–deforestation link may also exist. Finally, the evidence that deforestation may itself be an important cause of human-induced soil degradation across developing regions raises the possibility of a "cumulative causation" link between rural poverty, deforestation and land degradation. (Barbier 1997: 892)

In understanding the relationship between development and land degradation, the scale of analysis is an important factor. In addition to the usual factors highlighted in the literature, an additional aspect of the land degradation-development relationship in the context of the study area is the question of accessibility. Ravines are not just an indicator or type of land degradation; ravine formation at a scale that is observed in the Chambal region affects accessibility of the entire region. Thus, apart from the effects of soil health on farmland, there are broader, regional-level effects of ravines that have a strong impact on land degradation. In order to capture such interrelationship between land degradation and economic development at different spatial scales, an attempt has been made here to make a quantitative assessment of land degradation at the village level. This exercise is based on the analysis of the secondary data, mostly available in the Primary Census Abstract of the study area. Based on the integration of the land degradation map of Morena with the digitised location map of the villages of the district, we have categorised the villages into two groups: (i) those affected by the ravines and (ii) those not affected by the ravines. The land degradation-development relationship is then studied by a pairwise comparison of these two groups of villages across various development indicators. In case there is no relationship between these two, then the average levels of development of these two sets of villages should not vary significantly. However, if there is a systematic difference in the score of these two sets of villages in terms of their levels of development, it suggests the existence of a relationship between economic development of the village and its land degradation status.

4.2 Village-Level Characteristics of Morena

In the previous chapter, a socio-economic profile of the study region has been presented in a comparative framework. The purpose of the exercise was to locate the development parameters of the study region in comparison with the all-India and state averages. The exercise clearly brings out the relative backwardness of Morena in terms of many development indicators. The focus of this section is to analyse the

---

details, see Barbier, E.B., 2000. The economic linkages between rural poverty and land degradation: some evidence from Africa. *Agriculture, Ecosystems & Environment*, *82*(1–3), pp.355–370.

Table 4.1 Population size class-wise distribution of villages in blocks of Morena district

Population size	Ambah	Porsa	Morena	Pahadgarh	Joura	Kailaras	Sabalgarh	Total
Less than 200	2 (3%)	1 (1%)	4 (2%)	10 (8%)	0 (0%)	3 (3%)	6 (5%)	26 (3%)
200–499	4 (5%)	3 (4%)	21 (12%)	25 (20%)	7 (7%)	5 (5%)	18 (15%)	83 (11%)
500–999	7 (10%)	13 (18%)	35 (20%)	30 (24%)	21 (20%)	28 (27%)	35 (29%)	169 (22%)
1000–1999	19 (26%)	19 (27%)	53 (30%)	35 (28%)	39 (37%)	43 (41%)	35 (29%)	243 (31%)
2000–4999	28 (38%)	22 (31%)	50 (28%)	21 (17%)	31 (29%)	24 (23%)	20 (17%)	196 (25%)
5000–9999	11 (15%)	12 (17%)	12 (7%)	4 (3%)	7 (7%)	2 (2%)	5 (4%)	53 (7%)
1000 and above	2 (3%)	1 (1%)	1 (1%)	0 (0%)	1 (1%)	0 (0%)	0 (0%)	5 (1%)
Total	73	71	176	125	106	105	119	775

Source: Primary Census Abstract, Census of India 2011

important variations in levels of socio-economic indicators within the district, particularly focusing upon the distribution of villages according to the levels of socio-economic development. In the initial sections, the distribution of villages across different parameters provides the spatial dimensions of development in the region[6]. In the subsequent analysis, the differences among the land degradation-affected and non-affected villages have been compared in a similar framework. This exercise is an attempt to provide a context to the subsequent analysis based on primary data.

4.2.1 *Population Size*

The distribution of the villages according to population size reveals that nearly 31% of villages are in the size class of 1000–1999 and 22% are under the size class of 500–999, while 25% fall in the size class of 2000–4999 (Table 4.1). This pattern is more or less similar in all the C.D. blocks of the district. Notable exceptions include the following: (a) The share of relatively thinly populated villages (i.e. less than 500) is higher in Pahadgarh (28%) and Sabalgarh (20%); (ii) the share of relatively

[6] While the focus of the previous section was to compare the levels of socio-economic development in the district and tehsil levels, to bring out the differences within the district in a comparative framework, the analysis in this chapter goes a step further to bring in the interdistrict variations to the forefront. The exercise aims to capture the differences in the levels of development in the villages located in the degraded and non-degraded regions. The background information in Tables 4.1, 4.2, 4.3, 4.4, 4.5, 4.6, 4.7 and 4.8 helps to understand the nature of these differences.

bigger villages (more than 5000 population) is higher in Ambah (18%) and Porsa blocks (18%).

4.2.2 *Sex Ratio*

The region, particularly Morena, is known for its relatively low sex ratio. However, the distribution of the villages and populations according to the range of sex ratio reveals the severity of the problem. Nearly 5% of villages have a sex ratio of less than 700, and 12% of villages have a sex ratio of less than 750, with a population share of 5% (Table 4.2). This indicates that sex ratio is probably lower in villages with relatively small size of population. The adult sex ratio is also affected by the extent of sex-selective migration. But in general, the gender relations in the region are known to be less egalitarian, and hence the low sex ratio reflects female disadvantage rather than any other socio-economic phenomenon. The relationship between gender relations and environmental degradation has been highlighted in the literature (Leach et al. 1995). Land degradation as a specific form of environmental degradation is also known to have important gendered implications (Biot et al. 1995; Warren 2002).

Child sex ratio (CSR) is a better indicator to understand female disadvantage in a region. As high as 16. 3% of villages and 8% of population are in the lowest sex ratio class, i.e. below 700. Nearly 26.5% of villages have a CSR that is below 750 (Table 4.3). Compared to the national and state averages, this is a CSR that is deeply biased against the girl child.

Table 4.2 Distribution of villages and population by range of sex ratio, Morena in 2011

Range of sex ratio for villages	Number of inhabited villages	Percentage of villages in each range	Population 2011	Percentage distribution of population
Less than 700	42	5.42	28,509	1.91
700–749	49	6.32	56,275	3.76
750–799	124	16.00	177,140	11.84
800–849	252	32.52	550,782	36.83
850–899	227	29.29	593,468	39.68
900–949	64	8.26	76,202	5.10
950–999	12	1.55	11,942	0.80
1000–1099	5	0.65	1190	0.08
1100+	0	0.00	-	0.00
District: Morena (419)	**775**	**100.00**	**1,495,508**	**100.00**
Sex ratio district (Rural): 834				

Source: Primary Census Abstract, Census of India 2011

Table 4.3 Distribution of villages and population by range of child sex ratio, Morena in 2011

Range of sex ratio for villages	Number of inhabited villages	Percentage distribution of villages	Population 2011	Percentage distribution of population
Less than 700	126	16.26	18,682	7.81
700–749	77	9.94	23,299	9.74
750–799	92	11.87	32,496	13.58
800–849	142	18.32	67,623	28.26
850–899	114	14.71	46,967	19.63
900–949	88	11.35	25,859	10.81
950–999	38	4.90	9693	4.05
1000–1099	54	6.97	9881	4.13
1100+	44	5.68	4809	2.01
District: Morena (419)	775	100.00	239,309	100.00
Sex ratio district (Rural): 829				

Source: Primary Census Abstract, Census of India 2011

4.2.3 *Proportion of Scheduled Castes and Scheduled Tribes Population*

The proportion of scheduled caste and scheduled tribe is an important aspect of the social context in rural India. In 147 (19%) villages, there were no SC population, whereas in about 57 villages, the share of SCs was more than 50% (Table 4.4). The district has comparatively less share of ST population than that of the SC population. There was no ST population in 580 villages (75%). There were only 12 villages where the share of the ST population was more than 50% (Table 4.5).

4.2.4 *Literacy Rate*

While average literacy rate provides a summary picture of the average standards of education in a region, the distribution of villages according to the range of literacy rates provides clues to understand the spatial distribution of the literate population. In this case, the majority of villages and populations fall in the range of 51–80% of literacy rate. The outliers are important here. There were 16 villages where the literacy rates were in the range of 11 to 40%. On the other hand, there were 58 villages with a population share of 5% which had a literacy rate in the range of more than 80% (Table 4.6).

The distribution of villages according to range of literacy rates of the SC population brings out the distribution of literates among the socially marginalised population. As in the case of the general population, in the case of SCs as well, it was found that the majority of villages (86%) and populations were in the literacy

Table 4.4 Distribution of villages and population by share of SC population, Morena in 2011

Percentage range of scheduled castes population to total population	Number of villages	Percentage	Scheduled castes population	Percentage
NIL	147	18.97	0	0.00
Less than 5	89	11.48	4037	1.26
5–10	77	9.94	11,286	3.51
11–20	151	19.48	63,639	19.79
21–30	131	16.90	87,996	27.37
31–40	88	11.35	73,973	23.01
41–50	35	4.52	26,028	8.10
51–75	44	5.68	43,977	13.68
76 and above	13	1.68	10,594	3.29
District: Morena (419)	**775**	**100.00**	**321,530**	**100.00**

Source: Primary Census Abstract, Census of India 2011

Table 4.5 Distribution of villages and population by share of ST population, Morena in 2011

Percentage range of scheduled tribes population to total population	Number of villages	Percentage	Scheduled tribes population	Percentage
NIL	580	74.84	0	0.00
Less than 5	165	21.29	2221	15.46
5–10	1	0.13	51	0.36
11–20	5	0.65	2246	15.64
21–30	7	0.90	2843	19.80
31–40	4	0.52	1460	10.17
41–50	1	0.13	185	1.29
51–75	8	1.03	3762	26.19
76 and above	4	0.52	1594	11.10
District: Morena (419)	**775**	**100.00**	**14,362**	**100.00**

Source: Primary Census Abstract, Census of India 2011

rate range of 50–80%. However, there were 12 villages where the share of ST literates was less than 30% (Table 4.7).

So far as the distribution of villages according to the literacy rates of STs is concerned, only 24% of the villages fall under the range of 50–80% literacy. This suggests that not only the average ST literacy rate is lower than that of the whole population, even in comparison with the SCs, the literacy rate of STs is much worse. The distribution also suggests that in nearly 20% of ST-inhabited villages, there was not a single person who was literate. This shows extreme spatial and social exclusion of the scheduled tribes. In nearly 9% of villages, the literacy rate among the scheduled tribes was in between 11% and 40% (Table 4.8). Even in terms of such basic indicators, the structure of social deprivation in the study area comes out clearly.

This section provides an overview of the socio-economic characteristics of the villages in the study area. It is essential to understand the socio-economic context of

Table 4.6 Distribution of villages and population by literacy rate, Morena in 2011

Range of literacy rate for villages	Number of inhabited villages	Percentage distribution of villages	Population	Percentage distribution of population
0	0	0.00	0	0.00
1–10	0	0.00	0	0.00
11–20	1	0.13	383	0.03
21–30	3	0.39	1578	0.11
31–40	12	1.55	7244	0.48
41–50	24	3.10	17,420	1.16
51–60	151	19.48	225,423	15.07
61–70	284	36.65	578,885	38.71
71–80	242	31.23	587,021	39.25
81–90	51	6.58	73,984	4.95
91–99	6	0.77	3559	0.24
100	1	0.13	11	0.00
District: Morena (419)	**775**	**100.00**	**1,495,508**	**100.00**
Literacy rate for district:	**68.91**			

Source: Primary Census Abstract, Census of India 2011

the study region. The indicators discussed in the section, viz. population size, relative share of social groups, sex ratio, etc., are well recognised as key variables to understand the development process.[7] Two important insights that can be inferred from the analysis of the distribution of villages across size classes of different attributes are as follows: (i) as in the case of other fragile ecological regions, spatial distribution of the population across the villages is an important aspect of the study region; and (ii) the spatial pattern also gets reinforced by the linkages among the distribution of villages according to various indicators. For example, the correspondence between educational and social backwardness can be clearly seen from Tables 4.7 and 4.8.

[7] The indicators adopted for the analysis in this study are guided by two important considerations. Firstly, the vast literature on indicators suitable for understanding the complexities associated with land degradation has been used to select the relevant indicators; and secondly, the availability of information in the secondary data sources have also been taken into consideration while selecting the district, sub-district and village-level indicators. The indicators used in the study are similar to those used and recommended in a number of well-established studies on land degradation. For example, see Blaikie (1985: 30–3; 56–7); Svenson (2005); Nachtergaele and Licona-Manzur (2008); Sommer et al. (2011); Salvati and Zitti (2009); Nkonya et al. . (2011); Pani and Carling (2013); Mythili and Goedecke (2016).

Table 4.7 Distribution of villages and population by literacy rate (SC), Morena in 2011

Range of literacy rate for villages	Number of inhabited villages having scheduled castes	Percentage distribution of villages	Scheduled castes population	Percentage distribution of population
0	6	0.96	17	0.01
1–10	3	0.48	477	0.15
11–20	1	0.16	5	0.00
21–30	2	0.32	120	0.04
31–40	19	3.03	2175	0.68
41–50	32	5.10	5700	1.77
51–60	106	16.88	37,076	11.53
61–70	291	46.34	188,116	58.51
71–80	141	22.45	83,172	25.87
81–90	19	3.03	3977	1.24
91–99	4	0.64	664	0.21
100	4	0.64	31	0.01
Total	**628**	**100.00**	**321,530**	**100.00**
District scheduled castes literacy rate: 66.50				

Source: Primary Census Abstract, Census of India 2011

Table 4.8 Distribution of villages and population by literacy rate (ST), Morena in 2011

Range of literacy rate for villages	Number of inhabited villages having scheduled tribes	Percentage distribution of villages	Scheduled tribes population	Percentage distribution of population
0	38	19.49	91	0.63
1–10	0	0.00	0	0.00
11–20	2	1.03	39	0.27
21–30	6	3.08	297	2.07
31–40	10	5.13	1502	10.46
41–50	28	14.36	3840	26.74
51–60	22	11.28	4580	31.89
61–70	14	7.18	3465	24.13
71–80	10	5.13	264	1.84
81–90	7	3.59	85	0.59
91–99	1	0.51	24	0.17
100	57	29.23	175	1.22
	195	**100.00**	**14,362**	**100.00**
Literacy rate for district: 52.86				

Source: Primary Census Abstract, Census of India 2011

4.3 Broad Features of Rural Development

In the previous section, the discussions were about the outcome indicators, such as literacy and sex ratio. These measures of socio-economic development are influenced by the availability of infrastructure. In this section, a description of the distribution of the villages according to availability of some crucial indicators has been presented on the basis of the data available through the Primary Census Abstract. The choice of indicators is dependent on the availability of the village-level infrastructure data generated through the population census.

As per Table 4.9, in the district as a whole, 98% of villages have some education facilities, and only 6.3% have some medical facilities. Nearly 30% of villages have post offices, and 46% have any kind of transport facilities. Nearly 76% villages are connected through pucca road. Ambah and Porsa blocks are relatively better off in terms of availability of basic infrastructure. The geographical specificity of the region can best be judged from the fact that in the Morena community block that is supposed to be the nearest to the district headquarters, only 5.45% of villages have medical facility; only 9% of villages have banks; and only 36% of villages have communication facility of any kind. The villages located within the deep ravines are inaccessible, and infrastructure facilities do not reach these villages, irrespective of their location in the administrative location.

The distribution of population having access to different amenities, presented in Table 4.10, further strengthens the notion of spatial exclusion of a section of population. While a high share of population living in the rural areas of Morena have access to primary schools, power supply, drinking water, etc., access to medical facilities, banks, agricultural credit societies and transport and communication facilities is available to lesser percentages of the rural populations. The spatial disparities could be judged from the fact that the range of population having access to medical facilities is as low as 50% in Kailaras and as high as 83% in Porsa. What is even more surprising is that in Morena C.D. block, nearly 20% of population do not have access to pucca road, whereas in Ambah nearly 3% do not have such access.

Distance is a critical factor in accessibility, particularly in areas with a difficult terrain. The nature of spatial exclusion in the study region is further illustrated by the distribution of villages not having certain facilities by distance. For example, of all villages that do not have a primary school, 86% have a school within 5 km^2. But nearly 54% of villages not having a hospital in the village have to travel more than 10 kilometres to access the facility (Table 4.11). More strikingly, of those villages that do not have a bus facility in the village itself, 27% have to travel more than 10 kilometres to reach the bus service.

Access to amenities by distance to nearby towns is another important aspect of rural-urban interactions. Usually villages which are nearer to urban centres and those which are connected by pucca road have better access to amenities of various kinds. For most of the amenities, the share of villages having access to the amenities goes down with increasing range of distances from the urban centres (Table 4.12), suggesting an inverse relationship between distance and access to amenities. However, in a region characterised by dense ravines and undulated terrains, communication bottlenecks cannot always be measured through distance alone.

Population size also tends to influence levels of development for various reasons. Firstly, populations get concentrated around available amenities. Secondly, in a democratic set-up, population size is an important indicator of bargaining with the politicians and administrators at a local level. Thirdly, many of the programmes use population size as a criteria for allocation of funds and facilities. Thus, it is found

Table 4.9 Distribution of villages by availability of different amenities, Morena in 2011

Sr. No.	Name of C.D. block	Number of inhabited villages	Type of amenity available									
			Education[a]	Medical ^	Drinking water	Post office#	Telephone[b]	Transport communications $	Banks@	ACS+	Approach by pucca road	Power supply
1	Ambah	73	73	46	73	31	51	45	12	17	66	73
			(100)	(63.01)	(100)	(42.47)	(69.86)	(61.64)	(16.44)	(23.29)	(90.41)	(100)
2	Porsa	71	71	47	71	31	53	41	8	13	60	71
			(100)	(66.2)	(100)	(43.66)	(74.65)	(57.75)	(11.27)	(18.31)	(84.51)	(100)
3	Morena	176	173	80	176	44	140	63	16	25	128	142
			(98.3)	(5.45)	(100)	(25)	(79.55)	(35.8)	(9.09)	(14.2)	(72.73)	(80.68)
4	Pahadgarh	125	121	47	125	29	82	64	4	17	73	110
			(96.8)	(37.6)	(100)	(23.2)	(65.6)	(51.2)	(3.2)	(13.6)	(58.4)	10 (88)
5	Joura	106	106	49	106	31	92	51	6	28	84	105
			(100)	(46.23)	(100)	(29.25)	(86.79)	(48.11)	(5.66)	(26.42)	(79.25)	(99.06)
6	Kailaras	105	104	37	105	18	90	43	3	3	80	104
			(99.05)	(5.24)	(100)	(17.14)	(85.71)	(40.95)	(2.86)	(2.86)	(76.19)	(99.05)
7	Sabalgarh	119	114	53	119	43	93	53	3	5	96	117
			(95.8)	(44.54)	(100)	(36.13)	(78.15)	(44.54)	(2.52)	(4.2)	(80.67)	(98.32)
	Total	775	762	359	775	227	601	360	52	108	587	722
			(98.32)	(6.32)	(100)	(29.29)	(77.55)	(46.45)	(6.71)	(13.94)	(75.74)	(93.16)

Note: (i) [a]Education includes all education facilities. (ii) ^ Medical includes all medical facilities. (iii) # Post office includes post office, telegraph office and post and telegraph office. (iv) $ Transport communication includes bus service, rail facility and navigable waterways. (v) @ Bank includes commercial bank and cooperative bank. (vi) [b]Telephone includes telephone, PCO and mobile. (vii) + ACS agricultural credit societies

Source: Primary Census Abstract, Census of India 2011

Table 4.10 Number and percentage of rural population served by different amenities, 2011

Sr. No.	Name of C.D. block	Total population of inhabited villages	Type of amenity available									
			Education[a]	Medical^	Drinking water	Post office #	Tele phone[b]	Transport communications $	Banks@	ACS+	Approach by pucca road	Power supply
1	Ambah	2,11,512	2,11,512	1,68,728	2,11,512	1,39,329	1,68,689	1,65,528	55,917	68,294	2,06,106	2,11,512
			(100)	(79.77)	(100)	(65.87)	(79.75)	(78.26)	(26.44)	(32.29)	(97.44)	(100)
2	Porsa	1,91,485	1,91,485	1,58,816	1,91,485	1,31,961	1,61,809	1,43,072	41,354	64,388	1,82,217	1,91,485
			(100)	(82.94)	(100)	(68.91)	(84.5)	(74.72)	(21.6)	(33.63)	(95.16)	(100)
3	Morena	3,59,597	3,59,027	2,49,025	3,59,597	1,40,213	3,15,836	2,08,549	65,989	68,357	2,87,316	3,42,127
			(99.84)	(69.25)	(100)	(38.99)	(87.83)	(49.58)	(18.35)	(19.01)	(79.9)	(95.14)
4	Pahadgarh	1,65,125	1,64,685	97,284	1,65,125	62,637	1,29,984	1,22,282	13,548	32,574	1,34,407	1,59,070
			(99.73)	(58.92)	(100)	(37.93)	(78.72)	(74.05)	(8.2)	(19.73)	(81.4)	(96.33)
5	Joura	2,32,986	2,32,986	1,55,414	2,32,986	1,30,647	2,12,996	1,47,048	34,779	94,460	9499	2,32,524
			(100)	(66.71)	(100)	(56.08)	(91.42)	(63.11)	(14.93)	(40.54)	(89.92)	(99.8)
6	Kailaras	1,65,745	1,65,425	81,876	1,65,745	52,592	1,46,270	83,570	13,894	11,924	1,37,110	1,65,187
			(99.81	(49.4)	(100)	(31.73)	(88.25)	(50.42)	(8.38)	(7.19)	(82.72)	(99.66)
7	Sabalgarh	1,69,058	1,68,203	88,421	1,69,058	84,609	1,49,144	99,721	17,886	18,547	1,55,827	1,67,854
			(99.49)	(52.3)	(100)	(50.05)	(88.22)	(58.99)	(10.58)	(10.97)	(92.17)	(99.29)
	Total	14,95,508	14,93,323	9,99,564	14,95,508	7,41,988	12,84,728	9,69,770	2,43,367	3,58,544	13,12,482	1,469,759
			(99.85)	(66.84)	(100)	(49.61)	(85.91)	(64.85)	(16.27)	(23.97)	(87.76)	(98.28)

Note: (i) [a]Education includes all education facilities. (ii) ^ Medical includes all medical facilities. (iii) # Post office includes post office, telegraph office and post and telegraph office. (iv) $ Transport communication includes bus service, rail facility and navigable waterways. (v) @ Bank includes commercial bank and cooperative bank. (vi) [b]Telephone includes telephone, PCO and mobile. (vii) + ACS agricultural credit societies
Source: Primary Census Abstract, Census of India 2011

Table 4.11 Distribution of villages without amenities, according to distance ranges from nearest available facilities, 2011

Village not having the amenity of		Distance range of place from the villages where the amenity is available			
		Less than 5 km	5–10 km	10+ km	Total
1. Education					
(a) Primary school	No	12	2	0	14
	%	85.71	14.29	0.00	100
(b) Middle school	No	280	58	10	348
	%	80.46	16.67	2.87	100
(c) Degree college	No	448	99	219	766
	%	58.49	12.92	28.59	100
2. Medical					
(a) Hospital	No	134	217	414	765
	%	17.52	28.37	54.12	100
(b) PHC	No	113	299	343	755
	%	14.97	39.60	45.43	100
3. Post office	No	288	184	76	548
	%	52.55	33.58	13.87	100
4. Telephone	No	60	69	45	174
	%	34.48	39.66	25.86	100
5. Bus service	No	163	146	115	424
	%	38.44	34.43	27.12	100
6. Bank					
(a) Commercial bank	No	80	218	448	746
	%	10.72	29.22	60.05	100
(b) Cooperative bank	No	110	241	393	744
	%	14.78	32.39	52.82	100
7. ACS^{+}	No	141	265	261	667
	%	21.14	39.73	39.13	100

Note: (i) Degree college includes art, engineering and medicine. (ii) Hospital includes allopathic and alternative medicine. (iii) Post office includes post office, telegraph office and post and telegraph office. (iv) Telephone includes telephone, PCO and mobiles. (v) Bus includes private and public. (vi) + ACS agricultural credit societies
Source: Primary Census Abstract, Census of India 2011

that the share of villages having access to particular facilities goes up steadily as we move from lower size class-wise villages to higher-sized villages (Table 4.13). The descriptive analysis of distribution of villages according to availability of different amenities points to a pattern of rural development, where remoteness is an important axis of exclusion. Further, area under irrigation and literacy rates are found to be positively correlated at the village-level (Table 4.14).

Table 4.12 Distribution of villages according to distance ranges from nearest statutory towns and availability of different amenities, 2011

Distance range from the nearest statutory town (in Kms)	No/%	No. of inhabited villages in each range	Type of amenity available							
			Education[a]	Medical^	Post office #	Telephone[b]	Transport communications $	Banks@	ACS+	Approach by pucca road
Less than 5	No	49	47	26	12	42	18	0	5	42
	%		95.92	53.06	24.49	85.71	36.73	0	10.2	85.71
5–15	No	377	376	183	100	305	172	38	59	312
	%		99.73	48.54	26.53	80.9	45.62	10.08	15.65	82.76
16–50	No	292	286	130	102	215	149	13	40	200
	%		97.95	44.52	34.93	73.63	51.03	4.45	13.7	68.49
51+	No	11	10	5	3	10	4	0	0	3
	%		90.91	45.45	27.27	90.91	36.36	0	0	27.27
Unspecified	No	46	43	15	10	29	17	1	4	30
	%		93.48	32.61	21.74	63.04	36.96	2.17	8.7	65.22
Total	No	775	762	359	227	601	360	52	108	587
	%		98.32	46.32	29.29	77.55	46.45	6.71	13.94	75.74

Note: (i) [a]Education includes all education facilities. (ii) ^ Medical includes all medical facilities. (iii) # Post office includes post office, telegraph office and post and telegraph office. (iv) $ Transport communication includes bus service, rail facility and navigable waterways. (v) @ Bank includes commercial bank and cooperative bank.(vi) [b]Telephone includes telephone, PCO and mobile. (vii) + ACS agricultural credit societies

Source: Primary Census Abstract, Census of India 2011

Table 4.13 Distribution of villages according to size class of population and amenities available, 2011

Population range	Number/ percentage	Number of inhabited villages in each range	Type of amenity available									
			Education[a]	Medical^	Drinking water	Post office#	Tele phone[b]	Transport communications $	Banks@	ACS+	Approach by pucca road	Power supply
1–499	Number	109	96	24	109	17	65	26	0	6	51	83
	Percentage		88.07	22.02	100	15.6	59.63	23.85	0	5.5	46.79	76.15
500–999	Number	169	169	50	169	27	118	56	0	8	101	142
	Percentage		100	29.59	100	15.98	69.82	33.14	0	4.73	59.76	84.02
1000–1999	Number	243	243	99	243	38	188	102	11	29	196	243
	Percentage		100	40.74	100	15.64	77.37	41.98	4.53	11.93	80.66	100
2000–4999	Number	196	196	128	196	96	175	118	20	41	183	196
	Percentage		100	65.31	100	48.98	89.29	60.2	10.2	20.92	93.37	100
5000–9999	Number	53	53	53	53	45	50	53	19	21	51	53
	Percentage		100	100	100	84.91	94.34	100	35.85	39.62	96.23	100
10,000 +	Number	5	5	5	5	4	5	5	2	3	5	5
	Percentage		100	100	100	80	100	100	40	60	100	100
District total	Number	775	762	359	775	227	601	360	52	108	587	722
	Percentage		98.32	46.32	100	29.29	77.55	46.45	6.71	13.94	75.74	93.16

Note: (i)[a] Education includes all education facilities. (ii) ^ Medical includes all medical facilities. (iii) # Post office includes post office, telegraph office and post and telegraph office. (iv) $ Transport communication includes bus service, rail facility and navigable waterways. (v) @ Bank includes commercial bank and cooperative bank. (vi) [b]Telephone includes telephone, PCO and mobile. (vii) + ACS agricultural credit societies
Source: Primary Census Abstract, Census of India 2011

Table 4.14 Zero-order correlation matrix: village-level development indicators

	Literacy rate	Literacy rate males	Percentage of area irrigated	Share of non-farm male workers
Literacy rate	1	0.934**	0.237**	0
Literacy rate males		1	0.312**	0.003
Percentage of area irrigated			1	0.018
Non-farm				1

Source: Author's calculation based on Primary Census Abstract, Census of India 2011

4.4 Land Degradation and Socio-economic Development: Comparison Between Affected and Non-affected Villages

The availability of village-level statistics on the level and pattern of socio-economic development allows a comparison between degraded and non-degraded village. Following a novel methodology and combining census data with land use classification and ravine classification data, using 40 years of geospatial data analysis, this study presents a methodological innovation for interlinking socio-economic and environmental indicators. To understand the status of villages in terms of gully-affected villages and non-gully-affected villages, the village boundary of the study districts (total number of villages is 794 as per census) has been digitized using ArcGIS software from the district village boundary maps (district village boundary maps), and all the villages were superimposed on earlier prepared ravine maps of 2014. This analysis allowed further classification of villages on the basis of intensity of ravines such as low, moderate and highly degraded villages (see Fig. 2.4). For the comparison between villages with different levels of degradation, in this section, villages which are situated inside the gully-/ravine-affected area were classified as ravine-infested villages, and the rest of villages which are outside of the ravine area are classified as non-affected villages. In this process each village status has been identified. The nature of each village, their location, status and other socio-economic variables with information were added with the individual village and also evaluated using village census data of 2011 in GIS environment[8].

The comparison of villages which are grouped according to their land degradation status provides an understanding of the way land degradation and village-level socio-economic development are linked. However, before going for the analysis, Tables 4.15, 4.16, 4.17, 4.18, 4.19 and 4.20 provide a comparative picture of the status of these two sets of villages according to selected demographic characteristics. As the distribution of the villages into low, medium and high category of villages suggests (Table 4.15), most villages fall in the category of low levels of population. No systematic differences between degraded and non-degraded villages were found in the size class-wise distribution of villages (Table 4.16).

[8] For the details of methodology, see Sect. 1.13.

Table 4.15 Population wise distribution of villages: affected and non-affected villages

Category	Non-degraded villages	Degraded villages	All villages
Low	576 (94.4)	155 (93.9)	731 (94.3)
Medium	32 (5.2)	9 (5.5)	41 (5.3)
High	2 (0.3)	1 (0.6)	3 (0.4)
Total	610 (100)	165 (100)	775 (100)

Source: Author's calculations, based on Census of India 2011

Table 4.16 Distribution of population: Morena district, Madhya Pradesh

Size class of population	Degraded villages	Non-degraded villages	All villages
0–500	22 (13.3)	87 (14.3)	109 (14.1)
500–1000	30 (18.2)	139 (22.8)	169 (21.8)
1000–2000	52 (31.5)	191 (31.3)	243 (31.4)
2000–4000	41 (24.8)	127 (20.8)	168 (21.7)
4000–6000	11 (6.7)	41 (6.7)	52 (6.7)
More than 6000	9 (5.5)	25 (4.1)	34 (4.4)
Total	165 (100)	610 (100)	775 (100)

Source: Village Directory, Census of India 2011

Table 4.17 Distribution of villages according to levels of total literacy: affected and non-affected villages

Category	Non-degraded villages	Degraded villages	All villages
Low	12 (2.0)	4 (2.4)	16 (2.1)
Medium	432 (70.8)	121 (73.3)	553 (71.4)
High	166 (27.2)	40 (24.2)	206 (26.6)
Total	610 (100)	165 (100)	775 (100)

Source: Author's calculations, based on Census of India 2011

So far as literacy status is concerned, the share of low- and medium-literacy villages is higher among the degraded villages than in the non-degraded villages. The share of highly literate villages among the non-degraded villages is higher than that among the degraded villages (Table 4.17). The share of villages according to levels of male literacy depicts a similar picture (Table 4.18).

The share of irrigated area in net sown area is a reliable indicator of agricultural development. Availability of irrigation is generally associated with higher degree of adoption of modern technologies in agriculture. There are systematic evidences to

Table 4.18 Distribution of villages according to levels of male literacy

Category	Non-degraded villages	Degraded villages	All villages
Low	14 (2.3)	4 (2.4)	18 (2.3)
Medium	329 (53.9)	110 (66.7)	439 (56.6)
High	267 (43.8)	51 (30.9)	318 (41.0)
Total	610 (100)	165 (100)	775 (100)

Source: Author's calculations, based on Census of India 2011

Table 4.19 Distribution of villages according to share of irrigated area

Category	Non-degraded villages	Degraded villages	All villages
Low	89 (14.6)	46 (27.9)	135 (17.4)
Medium	179 (29.3)	70 (42.4)	249 (32.1)
High	342 (56.1)	49 (29.7)	391 (50.5)
Total	610 (100)	165 (100)	775 (100)

Source: Author's calculations, based on Census of India 2011

Table 4.20 Distribution of villages according to percentage of male non-farm workers

Category	Non-degraded villages	Degraded villages	All villages
Low	553 (90.7)	156 (94.5)	709 (91.5)
Medium	48 (7.9)	8 (4.8)	56 (7.2)
High	9 (1.5)	1 (0.6)	10 (1.3)
Total	610 (100)	165 (100)	775 (100)

Source: Author's calculations, based on Census of India 2011

show that earnings of farmers and levels of household welfare are higher in the irrigated villages than in the nonirrigated villages. The share of villages with low levels of irrigation is remarkably higher among the degraded villages than that among the non-degraded villages (Table 4.19). This shows that the linkage between development and land degradation probably works through its impacts on agricultural productivity. The share of male main workers engaged in nonagricultural activities (i.e. those other than cultivators and agricultural labourers) is considered as an indicator of the level of economic development[9]. Non-farm workers typically earn more than farm workers. Also, it is an indicator of diversification of the rural economy. The share of low levels of non-farm workers among the degraded villages is higher than that among the non-degraded village (Table 4.20).

4.4.1 Livelihoods in the Study Region

Livelihoods in the study region is predominantly agricultural. As it has been described in the previous chapter, a substantial section of the rural workers is dependent on agriculture, either as cultivators or as agricultural labourers. As reported in the previous chapter, in Morena district, in 2011, nearly 53% of the rural workers were cultivators, and 26% of rural workers were agricultural labourers. Thus, nearly 79% of rural workers were dependent on agriculture, suggesting a lack of alternative livelihoods options.

We computed village-level Livelihoods Diversification Index using the following formula:

$$LDI = 1 - \sum_{i=1}^{n} P_i^2$$

where P_i = proportion of workers in the i-th occupation to total workers.

The index takes the value 0, when all the workers are in the same occupation and higher levels of diversification are represented by higher values of the index. The purpose of estimating the Livelihoods Diversification Index is to examine the livelihood impacts of land degradation in the study region. Literature suggests that livelihoods can be diversified with respect to both expanding opportunities and lack of adequate earnings for survival. While the former is part of accumulative diversification, the latter is part of distress diversification. The nature of distribution of workers in different sectors (with different average levels of earnings) provides clues to understand the nature of diversification.

For 2001, the Livelihoods Diversification Index for Morena district works out to be 0.693, 0.638 and 0.507 for total, rural and urban workers, respectively. We have calculated the diversification index for all the 775 villages in the district separately (Table 4.21). The mean value of the LDI at the village level was found to be 0.402. However, the difference between mean value of the Livelihoods Diversification Index between degraded and non-degraded villages does not come out to be significant. The distribution of villages' Livelihoods Diversification Index shows that there are significant village-level differences in the extent of livelihoods diversification (Tables 4.22 and 4.23).

[9] Female workers are not included because the factors determining their participation in the non-farm economy are much more complex. It is possible, for example, that in a highly developed economy, participation of women workers in the non-farm economy is low because of social norms and restrictions.

Table 4.21 Distribution of villages according to share of non-farm workers in total workers

	Male		All workers	
Share of non-farm workers in total workers	No. of villages	Percent	No. of villages	Percent
Less than 5%	348	44.9	211	27.2
5–10%	172	22.2	179	23.1
10–20%	126	16.3	188	24.3
20–40%	86	11.1	139	17.9
40–60%	30	3.9	43	5.5
More than 60%	13	1.7	15	1.9
Total	775	100	775	100

Source: Author's calculations, based on Census of India 2011

Table 4.22 Mean Livelihoods Diversification Index

	Number of villages	Mean LDI
Degraded	165	0.397
Non-degraded	610	0.403
All	775	0.402

Source: Author's calculations, based on Census of India 2011

Table 4.23 Distribution of villages according to Livelihoods Diversification Index

	All villages		Non-degraded villages		Degraded villages	
Levels of diversification	Number	Percent	Number	Percent	Number	Percent
Very low (Less than 0.22)	153	19.7	125	20.5	28	17.0
Low (0.22–0.40)	180	23.2	131	21.5	49	29.7
Medium (0.40–0.58)	316	40.8	253	41.5	63	38.2
High (More than 0.58)	126	16.3	101	16.6	25	15.2
Total	775	100	610	100	165	100

Source: Author's calculations, based on Census of India 2011

4.4.2 Comparison of Overall Development

Development is a multidimensional process. Literacy is considered to be an important aspect of human development. Comparison between degraded and non-degraded villages, presented in Tables 4.24 and 4.25, presents a complex picture. Interestingly, the average population of the degraded villages is found to be more than that of the non-degraded villages. Mean levels of literacy – for male, female and all persons – are found to be higher in the non-degraded villages. Percentage share of SC and ST population is found to be higher in the non-degraded villages. Percentage of area irrigated is higher in the degraded villages than in the non-degraded villages. Mean distance from the nearby statutory town is higher in the

Table 4.24 Comparison of degraded and non-degraded villages: literacy levels

Category	N	Total population	Literacy all persons	Literacy male	Literacy female
Non-degraded	610	1896.22	67.37	80.62	51.17
Degraded	165	2053.4	65.86	78.25	50.81
Total	775	1929.69	67.05	80.11	51.1

Source: Author's calculations, based on Census of India 2011

Table 4.25 Comparison of means: development indicators of degraded and non-degraded villages

Category	N	Percentage of SC and ST population	Percentage of area irrigated	Distance to nearest statutory towns	Percentage share of male main workers in nonagriculture
Non-degraded	610	22.26	58.42	15.81	33.66
Degraded	165	17.59	73.73	13.36	33.27
Total	775	21.26	61.68	15.29	33.57

Source: Author's calculations, based on Census of India 2011

case of the non-degraded villages. The share of male main workers in nonagriculture does not vary significantly among the two categories. However when we calculated a composite index of infrastructure development[10], it was found that the mean index value for degraded villages (−0.0369) was less than that of the non-degraded villages (0.01).

4.5 Conclusion

The village as a unit of social and economic organisation provides a clue to understand the process of economic development. In order to understand the linkages between land degradation and socio-economic development, a two-step procedure was adopted in this study. As a first step, the distribution of the villages across size classes was analysed to bring out different aspects and levels of economic development in the study area. And then, in the next step, a systematic comparison was made between less or (non-)degraded villages and degraded villages. Two key

[10]The Infrastructure Development Index is the mean score of the Educational Infrastructure Index, Road Connectivity Index, Power Supply Index, Village Services Index, Information and Entertainment Index and Health Infrastructure Index. All these indices were constructed on the basis of data available from the Primary Census Abstract and the Village Directory. The availability of a particular infrastructure within the village was assigned the value 5; in case it is available within 5 kms from the village, the assigned value was 3; if it is available within a distance of 5–10 km, the assigned value was 2; and if it was available at a distance of more than 10 km, the assigned value was 1. The scores generated for each of the infrastructures were standardized, and then average value of the indices was arrived at by taking the arithmetic mean of the Z-scores of indicators.

insights that have been brought out from the analysis of the distribution of villages across size classes of different attributes are as follows. Firstly, as in the case of other fragile ecological regions, spatial distribution of the population across the villages is an important aspect of the pattern of socio-economic development in this region. Secondly, the spatial pattern also gets reinforced by the linkages among the distribution of villages according to various indicators.

The results show that in terms of some important indicators of development – such as levels of literacy and infrastructure availability – the degraded villages are lagging behind. However, the impact of land degradation is difficult to measure at the village-level data. Some of the villages affected by ravines, for example, are located near flood plains and hence have better quality of irrigation. Public service provisioning is often mediated through complex processes of bargaining and patron-client relationship. Thus, it is not that surprising that the disparities in the mean of indicators of development do not show a consistent pattern, when a comparison between degraded and less degraded villages is made. This, in turn, points to the significant role of infrastructure creation and public services provisioning in overcoming the locational disadvantages of the degraded villages.

References

Barbier EB (1997) The economic determinants of land degradation in developing countries. Philos Trans R Soc Lond B 352(1356):891–899

Barbier EB (2000) The economic linkages between rural poverty and land degradation: some evidence from Africa. Agric Ecosyst Environ 82(1–3):355–370

Biot Y, Blaikie PM, Jackson C, Palmer-Jones R (1995) Rethinking research on land degradation in developing countries. The World Bank, Washington, DC

Blaikie P (1985) The political economy of soil erosion in developing countries. London and New York: Longman

Census of India (2011) Census of India 2011, series – 24 Part Xii-A, village and town directory, 2011, directorate of census operations, Madhya Pradesh, Morena (Part A and B). Available at http://censusindia.gov.in/2011census/dchb/DCHB_A/23/2302_PART_A_DCHB_MORENA.pdfandhttp://censusindia.gov.in/2011census/dchb/2302_PART_B_DCHB_MORENA.pdf

Gallup JL, Sachs JD, Mellinger AD (1999) Geography and economic development. Int Reg Sci Rev 22(2):179–232

Jodha NS (1998) Degradation an alternative explanation and possible solutions. Econ Polit Wkly 33(36):2384–2390

Leach M, Joeks S, Green C (1995) Gender relations and environmental change. IDS Bull 26(1):1–8

Mythili G, Goedecke J (2016) Economics of land degradation in India. In: Economics of land degradation and improvement – a global assessment for sustainable development. Springer, Cham, pp 431–469

Nachtergaele FO, Licona-Manzur C (2008) The land degradation assessment in drylands (LADA) project: reflections on indicators for land degradation assessment. In: The future of drylands. Springer, Dordrecht, pp 327–348

Nkonya E, Winslow M, Reed MS, Mortimore M, Mirzabaev A (2011) Monitoring and assessing the influence of social, economic and policy factors on sustainable land management in Drylands. Land Degrad Dev 22(2):240–247. https://doi.org/10.1002/ldr.1048

Pani P, Carling P (2013) Land degradation and spatial vulnerabilities: a study of inter-village differences in Chambal Valley, India. Asian Geogr 30(1):65–79. https://doi.org/10.1080/10225706.2012.754775. Routledge

Salvati L, Zitti M (2009) Assessing the impact of ecological and economic factors on land degradation vulnerability through multiway analysis. Ecol Indic 9(2):357–363

Sommer S, Zucca C, Grainger A, Cherlet M, Zougmore R, Sokona Y et al (2011) Application of indicator systems for monitoring and assessment of desertification from national to global scales. Land Degrad Dev 22(2):184–197

Svenson L (2005) Socio-economic indicators for causes and consequences of land degradation. LADA technical paper FAO, pp 2–3

Warren A (2002) Land degradation is contextual. Land Degrad Dev 13(6):449–459

Chapter 5
Land Degradation and Rural Development: A Field-Based Analysis

Abstract The fifth chapter is based on a primary survey to analyse rural development and land degradation. In this chapter, the data from the physical survey and household analysis have been used to get the ground reality of the ravine-affected villages. While agriculture is the main source of livelihood for a majority of farmers, in terms of share of income, non-farm employment has started to play a major role. Along with participation in the local, rural non-farm economy, some villagers, particularly those belonging to the younger age groups, have started to migrate to urban areas as casual workers in the urban informal sector.

Keywords Rural development · Household survey · Focus group discussion (FGD) · Cropping pattern change · Crop productivity · Land Levelling. Implications of Land Levelling · Land Degradation Policy

5.1 Basic Socio-economic Features of Households

The study region is among the relatively backward areas of Madhya Pradesh. However, as the analysis of the secondary data reveals, the nature of this backwardness is such that often it is not adequately captured in the secondary data. While in some aspects, the indicators for the region or the Morena district are at par with the state averages, in some other aspects, it lags behind the state. The topography and ecological conditions of the district are such that pockets of backwardness and areas of relative prosperity coexist within the administrative boundary of the districts and tehsils.[1] The purpose of this chapter is to provide a broad understanding of the livelihood scenario in the region, through the analysis of household-level data generated

[1]Villages on the sides of the national highways or those with irrigation facilities are much developed than those without these attributes. Thus, a proper understanding of the backwardness can be studied by conducting a micro-level analysis of the development process.

P. Pani, *Land Degradation and Socio-Economic Development*, Advances in Asian Human-Environmental Research,
https://doi.org/10.1007/978-3-030-42074-1_5

through a field survey. Insights from the primary data have been used to probe the interrelationship between land degradation and rural development in the region.

5.1.1 Socio-economic Profile of the Households

The sample survey was conducted in nine villages in Morena district. The sampling method through which the households were selected has already been discussed in Chap. 1. In total, 474 households were selected for a detailed socio-economic survey based on a structured questionnaire. While some of the bigger villages, such as Ratan Basai, have a higher share of the sample households, others have a relatively smaller share. In total 21.5% of the sample households belong to the SC, 25.1% belong to the OBC, and the rest 53.4% belong to the other communities in the sample. The concentration of specific castes in some of the villages can be seen from the fact that the share of SCs is as high as 70% in one of the villages, Devgarh, while in some others, their presence is not so significant.[2] Taking all samples together, we find that the average year of schooling is only 5.53 years. It is highest in Baghchini[3] and is lower in Dhaniram ka Pura, Kishroli.[4] The average size of households is 7.35, indicating the demographic backwardness of the region. In two of the villages, it is higher than eight members per households. The average landholding size is 7.48 bigha per household. In Kishroli, it is 14.88 bigha, and in Nandpura it is 12.75 bigha, while in Rithona village, it is 4.81 bigha (Table 5.1.1).

The study region is known for female disadvantage of various kinds. The social norms in the region have created various barriers for women and the girl child. Although any inference regarding sex ratio on the basis of such small sample size is likely to lead in a biased result, the overall sex ratio and the child sex ratio have been presented in Table 5.1.2. It is important to note that in overall terms, the sex ratio is skewed against the women. Even if we disregard the intervillage differences as the sample size is too small, overall the observed sex ratio is 731 females per males. The child sex ratio is also very low at 780 girls per boys. However, since the child sex ratio is better than the overall sex ratio, it probably points to a course correction. Male-selective outmigration could also have caused such a pattern.

[2] These are sample characteristics and are not necessarily identical with the district or village characteristics. However, our sample households are representative in terms of the core issue under investigation, i.e. land degradation and agriculture.

[3] Bagchini is one of the prominent villages in this region, in spite of the fact that it falls inside the ravine area. This village, located at an advantageous situation, is joined with an important road. Moreover, this village and the surrounding areas got major attention in the early 1970s by the government and other agencies to develop coping mechanisms for land degradation. This village also has a long history of land levelling.

[4] The village Dhaniram ka Pura, Kishroli, is situated in a very remote location in the study region, very close to the bank of Chambal, and is connected through very highly dissected ravine areas, which are almost inaccessible during the rainy season, because its muddy road gets easily damaged by rain.

Table 5.1.1 Socio-economic profile of the sample households

Block	Villages	No of HH	Per cent to total sample HH	Social groups						Average years of schooling	Average HH Size	Average size of land-holding
				SC		OBC		Others				
				No	Per cent	No	Per cent	No	Per cent			
Ambah	Dhaniram ka Pura Kishroli	32	6.8	0	(0.0)	32	(100)	0	(0.0)	3.34	8.41	6.24
Ambah	Kishroli	48	10.1	2	(4.2)	31	(64.6)	15	(31.3)	4.27	8.46	14.88
Ambah	Dandoli	69	14.6	28	(40.6)	15	(21.7)	26	37.7)	5.7	7.43	6.3
Porsa	Ratan Basai	118	24.9	1	(0.8)	28	(23.7)	89	(75.4)	4.96	7.62	6.25
Ambah	Rithona	72	15.2	33	(45.8)	1	(1.4)	38	(52.8)	6.49	6.78	4.81
Jora	Baghchini	14	3.0	3	(21.4)	0	(0.0)	11	(78.6)	6.79	6.64	9.5
Jora	Devgarh	13	2.7	9	(69.2)	1	(7.7)	3	(23.1)	6.08	5.92	6.46
Ambah	Easha	94	19.8	25	(26.6)	10	(10.6)	59	(62.8)	6.47	6.73	7.7
Jora	Nandpura	14	3.0	1	(7.1)	1	(7.1)	12	(85.7)	6	7.43	12.75
	All	474	100	102	(21.5)	119	(25.1)	253	(53.4)	5.53	7.35	7.48

Sources: Field Survey

Note: (i) HH = households; (ii)∗ N = 468; (iii) only 2.1% female-headed households have been observed in all the study villages.(iv) The minimum year of schooling is 0 and the maximum year is 13 years

Table 5.1.2 Total population, sex ratio and child sex ratio

Village	No of individuals	Total children	Sex ratio	Child sex ratio (0–6)
DP Kishroli	269	118	781.46	866.67
Kishroli	394	143	569.72	920.00
Dandoli	503	209	710.88	743.59
Ratan Basai	889	383	756.92	868.13
Rithona	484	205	734.77	666.67
Baghchini	92	39	735.85	1000.00
Devgarh	76	27	551.02	444.44
Easha	636	289	832.85	724.64
Nandpura	104	43	704.92	444.44
Total	3447	1456	731.29	780.19

Source: Field Survey

In order to understand the socio-economic conditions of the households, it is useful to find out the principal occupation of the households. As per the data presented in Table 5.2, 81% of households are dependent on cultivation as their principal source of income, while nearly 9% depend on casual work in non-farm sector. Nearly 6% of household heads report regular salaried work as their principal occupation. Village wise, in five of the villages, the share of those dependent on agriculture is more than 80% (Table 5.2). This is partly the result of the sampling design, but it also indicates the high levels of dependence on agriculture.

5.2 Access to Land

Land is among the primary resources in a rural, agricultural economy. Considering all sample households, it is found that 1.30% are landless,[5] 13.30% are marginal farmers, 42.40% are small farmers and 26.20% are semi-medium farmers. This categorisation of farmers has been done on the basis of the prevailing patterns of landholdings in the study area, and hence, this is different from the standard definitions of different categories of farmers that followed in secondary data sources like the Agricultural Census and NSSO.

In most of the villages, small farmer category has a large share. However, village-specific differences in the share of different categories of ownership holdings are also significant. For example, 44% of sample households in Kishroli and 21% of sample households in Nandpura are in the large farmer category[6] with ownership of

[5] While the broad characteristics of land distribution, captured through the primary survey (viz. the preponderance of small farmers in the region, are consistent with the overall agrarian structure of the district, such a low level of landlessness needs to be understood in the backdrop of the sampling design adopted for the study.

[6] These two villages are situated on the flood plain of river Chambal but surrounded by the dense and severe matured ravine (Plates 5.1a, 5.1b and 5.2).

Table 5.2 Occupational distribution of households: principal occupation of household heads

Villages		Cultivators	Agricultural labour	Animal husbandry	Casual non-farm work	Regular salaried employee	Pensioner	Total
DP Kishroli[a]	N	28	0	0	2	2	0	32
	%	87.50	0.00	0.00	6.30	6.30	0.00	100.00
Kishroli	N	45	1	0	1	1	0	48
	%	93.80	2.10	0.00	2.10	2.10	0.00	100.00
Dandoli	N	60	1	1	0	5	2	69
	%	87.00	1.40	1.40	0.00	7.20	2.90	100.00
Ratan Basai	N	90	0	2	16	10	0	118
	%	76.30	0.00	1.70	13.60	8.50	0.00	100.00
Rithona	N	52	1	2	12	3	2	72
	%	72.20	1.40	2.80	16.70	4.20	2.80	100.00
Baghchini	N	9	0	0	4	0	1	14
	%	64.30	0.00	0.00	28.60	0.00	7.10	100.00
Devgarh	N	10	2	0	1	0	0	13
	%	76.90	15.40	0.00	7.70	0.00	0.00	100.00
Easha	N	76	2	1	6	6	3	94
	%	80.90	2.10	1.10	6.40	6.40	3.20	100.00
Nandpura	N	14	0	0	0	0	0	14
	%	100.00	0.00	0.00	0.00	0.00	0.00	100.00
Total	N	384	7	6	42	27	8	474
	%	81.00	1.50	1.30	8.90	5.70	1.70	100.00

Source: Field Survey
Note: [a]The complete name of the village *DP Kishroli* is *Dhaniram ka Pura Kishroli*

more than 14 bigha of land (Table 5.3). The overall picture that emerges from the land ownership data is that the region is characterised by the predominance of small holders. Nearly 20% of land-holders do not own their land. They either have leased-in land from someone else or have encroached upon the land. Similarly, nearly 20% of households have reported that they have leased-out their land to someone else (Table 5.4).

So far as irrigation status is concerned, nearly 48% of the farmers cultivate unirrigated land. Among those households who have access to irrigated land, 76% are dependent on wells and 23% are dependent on canal irrigation (Table 5.5). Nearly 99% of households own their homestead land. While 41% of households have reported no change of land ownership in the last 5 years, 10% of households reported an increase in the amount of land owned. In the majority of households, 57% reported a decline in land owned. Again, a substantial majority reported land degradation to be the cause behind the decrease in land owned (Table 5.6).

Plate 5.1a Pipe fitted to drain the overflowing water during rainy season on the road and prevent the road from further damaging

Plate 5.1b Pipe fitted in the levelled ravine land use for agriculture, to drain extra rainwater from the field, planted castor oil plant to check rain drops. (Source: Field Photographs from author's field work)

Plate 5.2 Earthen mound created to check runoff on agricultural field

Table 5.3 Land distribution among the study villages

Name of the study villages	No/ percentage	Size classes of land-holding						
		Landless (0.0) bigha	Marginal (0.1–2) bigha	Small (2.1–6) bigha	Semi-medium (6.1–10) bigha	Medium (10.1–14) bigha	Large (above 14.1) bigha	All
DP Kishroli	N	1	7	10	9	3	2	32
	%	3.10	21.90	31.30	28.10	9.40	6.30	100.00
Kishroli	N	1	2	10	12	2	21	48
	%	2.10	4.20	20.80	25.00	4.20	43.80	100.00
Dandoli	N	0	11	36	14	5	3	69
	%	0.00	15.90	52.20	20.30	7.20	4.30	100.00
Ratan Basai	N	1	8	67	32	4	6	118
	%	0.80	6.80	56.80	27.10	3.40	5.10	100.00
Rithona	N	0	15	44	8	5	0	72
	%	0.00	20.80	61.10	11.10	6.90	0.00	100.00
Baghchini	N	0	3	1	6	1	3	14
	%	0.00	21.40	7.10	42.90	7.10	21.40	100.00
Devgarh	N	0	3	4	4	2	0	13
	%	0.00	23.10	30.80	30.80	15.40	0.00	100.00
Easha	N	3	13	27	33	13	5	94
	%	3.20	13.80	28.70	35.10	13.80	5.30	100.00
Nandpura	N	0	1	2	6	2	3	14
	%	0.00	7.10	14.30	42.90	14.30	21.40	100.00
All	N	6	63	201	124	37	43	474
	%	1.30	13.30	42.40	26.20	7.80	9.10	100.00

Source: Field Survey

Table 5.4 Tenancy status of land

Category	Response	Number	Percentage
Land lease-in or encroached	No	380	80.2
	Yes	94	19.8
Land lease-out or uncultivated	No	378	79.7
	Yes	96	20.3
Total observations	Total	474	100

Source: Field Survey

Table 5.5 Sources of irrigation

Variables		No. of observations	Percentage
Sources of irrigation	Tank	3	0.63 [1.23]
	Well	185	39.03 [76.13]
	Canal	55	11.60 [22.63]
	Rainfed	227	47.89
	Non-response	4	0.84
	All	474	100.00

Sources: Field Survey
Note: Figures within square brackets refer to percentage to total households reporting irrigated land (243)

Table 5.6 Changes in land ownership

Variables	Response	No. of observations	Percentage
Own homestead land	Yes	470	99.2
	No	4	0.8
Change in land ownership in the last 5 years	Increased	10	2.1
	Decreased	272	57.4
	Unchanged	192	40.5
Land ownership changed (decreased) due to land degradation	Yes	235	86.4
	No	37	13.6
Total	*N*	272	100

Sources: Field Survey

5.3 Crop Production and Productivity

As already mentioned, crop cultivation is the primary source of livelihood for the majority of the sample households. The number of households reporting cultivation of different crops and the respective levels of crop productivity has been reported in Table 5.7. Nearly all households are engaged in the production of wheat,[7] while a

[7]Although the preferred crops are mustered in unirrigated area and levelled area, some of the surveyed villages are located in flood plain; therefore households are engaged in the production of wheat due to plenty of water available for those lands.

Table 5.7 Crop production and productivity

	Households engaged in crop production		Productivity per bigha (qntls./bigha)*
Crops	No	Percentage	
Wheat	473	99.8	7.35
Maize	31	6.5	3.19
Bajra	432	91.1	4.55
Rice	22	4.6	2.06
Mustard	391	82.5	2.26
Soan	58	12.2	2.18
Til	10	2.1	0.91
Arhar	3	0.6	1.80
Potato	4	0.8	41.74
Sugarcane	1	0.2	n.a.
Fodder crops	32	6.8	n.a.
Total	474		

Sources: Field Survey

Note: Data on sugarcane and fodder production not reported because of small number of observation or non-reporting

* Bigha is the local unit of measurement

significant majority also cultivate bajra and mustard. Thus the crop production system of the study region is primarily geared towards these crops. Major crops of this area are wheat or mustard and bajra (pearl millet).[8] Wheat and bajra cycle is subject to the availability of irrigation. But in most of the cases, mustard is the preferred crop in the levelled lands, where no irrigation facility is available, as reported by individual farmers of different villages and also at the time of the FGD. The altered gully beds are also used as a farmland (Plate 5.3) in the study region to cultivate millet.

Many villagers who are involved in farming reported[9] that rain plays an important role as a deciding factor of the choice of crop. Generally the engineered gully or ravine beds and levelled lands are used for single cropping, but if the area receives rain after the millet harvesting, then they grow mustard in that year in the same field; otherwise the land is left as a fallow land for rest of the year (Plates 5.4a and 5.4b). But almost each and every farmer agreed that the first time after the levelling, the land is very productive, but productivity declines in later years. However, almost every farmer has shown reluctance to use fertiliser in terraced *behad* farmland and levelled land. Apart from manual labour, they do not invest any input in *behad zamin* (ravine-related land). Crop damage by wild animals has also been a matter of concern by the farmers. There has been a rise in human-wildlife conflict in recent years. Due to migration and lack of manpower to protect the farm, farmers build permanent houses in the agricultural land and prefer a short-term crop rather than *arhar* (lentils), *chole* (pigeon pea), etc. which are long-standing crops and difficult

[8] Botanical name is *Pennisetum glaucum*.

[9] Interview with a primary school teacher, age 49, in Gospura village, small farmer and part-time barber, age 35, an agriculture labourer, age 45, and a large farmer, aged 70.

Plate 5.3 An unsuccessful attempt to prevent runoff on the agricultural field using earthen bund; arrow is showing the damaged area of the bund. (Source: Field Photograph from author's field work)

Plate 5.4a Cultivation on recently levelled land, here, looks like a rolling topography

Plate 5.4b Deep ravine land levelled for agricultural use. (Source: Field Photograph from author's field work)

to protect from the wild animals *like nilgai* (antelope) or wild boar. Flocks of peacocks bring damage to pigeon pea; due to this, it is grown less for the last few years.[10]

5.4 Livelihood Diversification

Livelihood diversification has been identified as a key aspect of rural economic transformation. It is defined as 'the process by which rural families construct a diverse portfolio of activities and social support capabilities in order to survive and to improve their standards of living' (Ellis et al. 1993: 4). Thus, livelihood diversification can be for survival or distress-driven, and it can also be a response to opportunities for improving earnings or assets. In fragile ecologies, poor households typically opt for less risky options. One of the ways through which they minimise their risks is by spreading their assets, activities and skills on a number of different activities. Within agriculture, it might include a diversified cropping pattern. In terms of employment opportunities, it involves distribution of household labour in different activities. Livelihood diversification also involves migration[11] or commuting by some members to other areas.

[10] As per discussions during FGD with farmers

[11] Further details are placed in migration section.

Table 5.8 Livestock ownership

Village	Cattle per household (Nos)	Standardised livestock units (SLU[a])
DP Kishroli	3.21	3.70
Kishroli	4.64	5.24
Dandoli	3.26	3.56
Ratan Basai	2.59	3.03
Rithona	2.75	2.88
Baghchini	2.42	2.29
Devgarh	2.73	2.40
Easha	2.85	3.04
Nandpura	2.71	2.88
Total	3.01	3.31

Note: [a]SLU has been calculated with the following weights: cow 1,bullock 1,donkey 0.7, horse and pony 1, goat, sheep and lamb 0.1, poultry =0.033

5.4.1 *Livestock*

In fragile ecological zones, livestock rearing provide an alternative source of earnings, particularly for the poor. Participation of women and children in livestock rearing is often found to be significant. Livestock economy is also significant for agriculture. Animal power is used for agricultural operations, and also, it is a source of manure. The significance of livestock for the livelihoods of the surveyed households can be judged from the fact that only 5.1% of the households does not have any cattle. The average number of cattle owned per household was 3.01, and the average number of standardized livestock unit per household was 3.31 (Table 5.8). During field work, the villagers, particularly small-scale farmers, informed that there is a decline in the area of grazing land, as ravine areas are getting levelled and access to the common property is getting reduced. This high rate of privatisation of the common land is a matter of concern for the marginalised sections of the population in the area.[12]

5.4.2 *Livelihood Diversifications*

5.4.2.1 Participation in the Non-farm Economy

Generally, livelihood diversification in an agricultural economy takes place through participation in the non-farm economy. Participation in the non-farm economy could be through the local non-farm economy (also called rural non-farm economy) or through migration and commuting (Table 5.9.1).

[12]The decline in grazing land availability is reported to be most severe in the village Bagchini.

Table 5.9.1 Share of non-farm income among the study villages

Villages	Number/ percentage	Percentage of household having non-farm income	Percentage of households having more than 50% of income from non-farm sources
DP Kishroli	N	11	9
	%	34.40	28.10
Kishroli	N	18	12
	%	37.50	25.00
Dandoli	N	44	36
	%	63.80	52.20
Ratan Basai	N	68	62
	%	57.60	52.50
Rithona	N	50	49
	%	69.40	68.10
Baghchini	N	8	6
	%	57.10	42.90
Devgarh	N	9	9
	%	69.20	69.20
Esah	N	42	38
	%	44.70	40.40
Nandpura	N	3	3
	%	21.40	21.40
Total	N	253	224
	%	53.40	47.30

Sources: Field Survey
Note: The income data is based on reported income

The percentage of households reporting some non-farm income is as high as 53.4%, considering the data from all villages. If we take into account households with more than 50% of their income from nonagricultural sources, the share is 47.30%. This shows the significant role of the non-farm income and employment in the rural economy, even though we have considered only farmer households. In some villages, such as Rithona, Devgarh and Dandoli, the share of households reporting income from non-farm sources is relatively high. However, this is a tendency that has been confirmed by other studies from different parts of India.

5.4.2.2 Occupational Diversification at the Household Level

The livelihood diversification approach is used to understand the responses of households under ecological stress. The relationship between ecological stress and livelihood diversification is far from straightforward as there are many intervening variables that might affect the relationship between these two variables. For example, higher degree of soil fertility often leads to the emergence of a prosperous agricultural economy, which in turn leads to a process of agriculture-led livelihood diversification. Economist John Mellor and others have identified this as a path of agriculture-based rural development strategy. Often, decline in soil fertility, deforestation and other kinds of environmental destruction lead to a decline in the earnings from traditional occupations such as agriculture and animal husbandry. This induces a process of distress livelihood diversification. It is difficult to identify these processes as there may be other exogenous factors such as urbanisation that might create a demand for nonagricultural livelihoods, which is not dependent on the natural resource base of a particular rural region. Also, in any given context, there is unequal distribution of livelihood assets among households and individuals. Depending on their different resource endowment positions, households and individuals might react differently to environmental degradation.[13]

In the livelihood diversification literature, it is often emphasized that households rather than individuals should be taken as the unit of analysis while analysing livelihood diversification in developing economies. Further, the objective of the households may not always be the maximisation of household income; minimisation and dispersion of risks can also be an important objective of the households.

An analysis of occupational diversification has already been presented in Sect. 4.4.1, on the basis of the census data. The extent of household-level livelihood diversification in the study villages has been presented in Table 5.9.2. This analysis is based on the distribution of households according to the principal occupation of the head of the households. This suffers from the following limitations: (i) firstly, the occupation of all the adult members has not been taken into account in this exercise; and (ii) secondly, the analysis covers the principal occupations. In many developing economies, particularly in the context of fragile ecologies, households tend to depend on a number of part-time livelihoods. Such activities are generally undertaken for a few days in the year and are generally not captured by the data on principal occupation. Nevertheless, the analysis presented in Table 5.9.2, suggests that agriculture remains the topmost livelihood activity in the region. The extent of dependence on agriculture varies from a low of 64% to a high of 100%. Combining all surveyed households, it is found that 81% of the household heads have reported agriculture as the primary occupation. There are important village-level variations. But overall, casual non-farm wage employment and regular salaried occupation emerge as prominent categories of non-farm employment. Among the surveyed households, agricultural wage labour has a remarkably low presence as a principal occupation. It is partly caused by the nature of the agricultural economy in the region. It is a primarily rainfed agricultural belt, and hence agricultural work is not available throughout the year. Secondly, because our sample design was based on the selection of cultivator households, there is a higher observed dependence on agriculture.

[13] For example, in a village facing severe land degradation, individuals with poor human capital might be forced to opt for distress outmigration as a survival strategy, while individuals with better education or financial resources might opt for an accumulative diversification through outmigration.

Table 5.9.2 Distribution of households according to principal occupation of the head of the household

Occupation		DP Kishroli	Kishroli	Dandoli	Ratan Basai	Rithona	Bagchini	Devgarh	Esah	Nandpura	Total
Cultivators	N	28	45	60	90	52	9	10	76	14	384
	%	87.50	93.80	87.00	76.30	72.20	64.30	76.90	80.90	100.00	81.00
Agricultural labour	N	0	1	1	0	1	0	2	2	0	7
	%	0.00	2.10	1.40	0.00	1.40	0.00	15.40	2.10	0.00	1.50
Livestock	N	0	0	1	2	2	0	0	1	0	6
	%	0.00	0.00	1.40	1.70	2.80	0.00	0.00	1.10	0.00	1.30
Casual non-farm	N	2	1	0	16	12	4	1	6	0	42
	%	6.30	2.10	0.00	13.60	16.70	28.60	7.70	6.40	0.00	8.90
Regular salaried	N	2	1	5	10	3	0	0	6	0	27
	%	6.30	2.10	7.20	8.50	4.20	0.00	0.00	6.40	0.00	5.70
Pension and transfers	N	0	0	2	0	2	1	0	3	0	8
	%	0.00	0.00	2.90	0.00	2.80	7.10	0.00	3.20	0.00	1.70
Total	N	32	48	69	118	72	14	13	94	14	474
	%	100.00	100.00	100.00	100.00	100.00	100.00	100.00	100.00	100.00	100.00

Source: Field Survey

5.4.2.3 Occupational Diversification at the Individual Level

The distribution of individuals who are 15 years or above is presented in Table 5.9.3. As expected, there is a greater degree of diversification that is captured through the distribution of individuals across different occupations. In the total sample, nearly 41% of adults are in cultivation, and the second major occupational category is domestic work. The share of casual non-farm wage labour and salaried employment is much less than these two categories. The category domestic work is often considered as a non-working category, as it includes household chores and care activities. But in the study area, it was observed that women who identify themselves as home workers often combine a range of economic activities such as gathering food and fodder, animal husbandry and agricultural operations along with their reproductive burden.

This distribution of the adults is primarily based on the main economic activity, and it might under-represent the extent of diversification. Also, because the sampling design focused on selecting the farmer households, the extent of diversification captured by the data below can only be thought of as the extent of diversification among the cultivator households. The extent of diversification in the study villages is expected to be higher. Also, the seasonal migration activities, when these are only for a few weeks in a year, could have been under-reported in the survey. To overcome this lacuna, a separate analysis of the migrant workers has also been presented in this book. The occupational diversification of the adults is broadly consistent with the workforce structure data generated from alternative, secondary sources. Agriculture, particularly crop cultivation, remains the main occupation in the study region.

The gender division of work is an important feature of the social structure in the rural areas. The gender differences in employment and economic activities are captured in the analysis of occupational distribution of male and female adults in Tables 5.9.4 and 5.9.5, respectively. Among the male adults, 63% reported agriculture as their main occupation, 16% were students, and 13% were engaged in casual wage work in the non-farm sector. There are significant intervillage differences as well. But among the females, 79% reported themselves as domestic workers, and only 9% identified themselves as cultivators. The gendered construction of work and economic activities is clear from the differences noticed in the occupational distribution of men and women. This again is consistent with the observed low work participation rate among the females in the study region.

5.4.2.4 Livelihood Diversification Index

The extent of livelihood diversification in the sample households has been presented in Table 5.9.6. The diversification index used here is the Herfindahl index of diversification. While HID_1 refers to diversification index calculated on the basis of the main occupation of the household, HID_2 refers to diversification index calculated at the individual level. HID_2 has been computed separately for males, females and all

Table 5.9.3 Occupational distribution of workers

Villages		Cultivator	Agri-lab	Animal husbandry	Casual non-farm	Salaried service	Pension	Remittances	Students	Domestic work	Unemployed	Total
DP Kishroli	N	104	1	0	1	0	0	0	23	27	5	161
	%	64.60	0.60	0.00	0.60	0.00	0.00	0.00	14.30	16.80	3.10	100.00
Kishroli	N	150	2	0	5	5	1	0	32	81	4	280
	%	53.60	0.70	0.00	1.80	1.80	0.40	0.00	11.40	28.90	1.40	100.00
Dandoli	N	111	10	0	35	16	3	1	56	120	4	356
		31.20	2.80	0.00	9.80	4.50	0.80	0.30	15.70	33.70	1.10	100.00
Ratan Basai	N	206	0	1	47	10	2	0	85	196	3	550
	%	37.50	0.00	0.20	8.50	1.80	0.40	0.00	15.50	35.60	0.50	100.00
Rithona	N	124	2	0	41	11	1	0	46	103	3	331
	%	37.50	0.60	0.00	12.40	3.30	0.30	0.00	13.90	31.10	0.90	100.00
Bagchini	N	21	0	0	12	3	2	0	3	23	1	65
	%	32.30	0.00	0.00	18.50	4.60	3.10	0.00	4.60	35.40	1.50	100.00
Devgarh	N	17	2	1	10	1	0	0	4	18	0	53
	%	32.10	3.80	1.90	18.90	1.90	0.00	0.00	7.50	34.00	0.00	100.00
Esah	N	172	7	1	22	4	6	0	61	163	1	437
	%	39.40	1.60	0.20	5.00	0.90	1.40	0.00	14.00	37.30	0.20	100.00
Nandpura	N	33	0	0	2	1	0	0	8	24	0	68
	%	48.50	0.00	0.00	2.90	1.50	0.00	0.00	11.80	35.30	0.00	100.00
Total	N	938	24	3	175	51	15	1	318	755	21	2301
	%	40.80	1.00	0.10	7.60	2.20	0.70	0.00	13.80	32.80	0.90	100.00

Notes: Only 15+ population has been considered
Source: Field Survey

Table 5.9.4 Occupational diversification of males: 15+ age group

Villages		Cultivator	Agri-lab	Animal husbandry	Casual non-farm	Salaried service	Pension	Remittances	Students	Unemployed	Total
DP Kishroli	N	77	1	0	1	0	0	0	12	2	93
		82.80	1.10	0.00	1.10	0.00	0.00	0.00	12.90	2.20	100.00
Kishroli	N	146	2	0	5	4	1	0	28	1	187
		78.10	1.10	0.00	2.70	2.10	0.50	0.00	15.00	0.50	100.00
Dandoli	N	108	10	0	35	16	3	1	40	2	215
		50.20	4.70	0.00	16.30	7.40	1.40	0.50	18.60	0.90	100.00
Ratan Basai	N	194	0	1	46	10	2	0	57	2	312
		62.20	0.00	0.30	14.70	3.20	0.60	0.00	18.30	0.60	100.00
Rithona	N	103	2	0	41	11	1	0	34	2	194
		53.10	1.00	0.00	21.10	5.70	0.50	0.00	17.50	1.00	100.00
Bagchini	N	21	0	0	12	2	2	0	2	0	39
		53.80	0.00	0.00	30.80	5.10	5.10	0.00	5.10	0.00	100.00
Devgarh	N	17	2	1	10	1	0	0	3	0	34
		50.00	5.90	2.90	29.40	2.90	0.00	0.00	8.80	0.00	100.00
Esah	N	156	7	1	22	3	6	0	41	1	237
		65.80	3.00	0.40	9.30	1.30	2.50	0.00	17.30	0.40	100.00
Nandpura	N	31	0	0	2	1	0	0	3	0	37
		83.80	0.00	0.00	5.40	2.70	0.00	0.00	8.10	0.00	100.00
Total	N	853	24	3	174	48	15	1	220	10	1348
		63.30	1.80	0.20	12.90	3.60	1.10	0.10	16.30	0.70	100.00

Notes: Only 15+ population has been considered
Source: Field Survey

Table 5.9.5 Occupational diversification of females: 15+ age group

Villages		Cultivator	Casual non-farm	Salaried service	Students	Domestic work	Unemployed	Total
DP Kishroli	N	27	0	0	11	27	3	68
	%	39.70	0.00	0.00	16.20	39.70	4.40	100.00
Kishroli	N	4	0	1	4	81	3	93
	%	4.30	0.00	1.10	4.30	87.10	3.20	100.00
Dandoli	N	3	0	0	16	120	2	141
	%	2.10	0.00	0.00	11.30	85.10	1.40	100.00
Ratan Basai	N	12	1	0	28	196	1	238
	%	5.00	0.40	0.00	11.80	82.40	0.40	100.00
Rithona	N	21	0	0	12	103	1	137
	%	15.30	0.00	0.00	8.80	75.20	0.70	100.00
Bagchini	N	0	0	1	1	23	1	26
	%	0.00	0.00	3.80	3.80	88.50	3.80	100.00
Devgarh	N	0	0	0	1	18	0	19
	%	0.00	0.00	0.00	5.30	94.70	0.00	100.00
Esah	N	16	0	1	20	163	0	200
	%	8.00	0.00	0.50	10.00	81.50	0.00	100.00
Nandpura	N	2	0	0	5	24	0	31
	%	6.50	0.00	0.00	16.10	77.40	0.00	100.00
Total	N	85	1	3	98	755	11	953
	%	8.90	0.10	0.30	10.30	79.20	1.20	100.00

Notes: Only 15+ population has been considered
Source: Field Survey

individuals. It is found that there are considerable intervillage differences in the extent of livelihood diversification, as measured through the Herfindahl index. Overall, the extent of diversification is low, and most of the households and male workers are dependent on crop cultivation as their main occupation.

Although there are intervillage differences, by and large, livelihood diversification is higher among the males than that among the females. Given the low female work participation rates in the region, the results seem to be consistent with other studies regarding the region. Nevertheless, caution has to be exercised in interpreting and generalising the results presented here. The households were selected to capture the specific interrelationship between land degradation and agriculture; thus these may not represent the average levels of diversification in the study villages.

Table 5.9.6 Herfindahl index of diversification

	No of households	HID_1	HID_2M	HID_2F	HID_2T
DP Kishroli	32	0.227	0.050	0.500	0.347
Kishroli	48	0.120	0.144	0.111	0.511
Dandoli	69	0.237	0.557	0.048	0.677
Ratan Basai	118	0.392	0.377	0.117	0.610
Rithona	72	0.447	0.503	0.281	0.651
Baghchini	14	0.500	0.567	0.080	0.697
Devgarh	13	0.379	0.589	0.000	0.701
Esah	94	0.337	0.345	0.172	0.597
Nandpura	14	0.000	0.164	0.142	0.536
All	474	0.332	0.391	0.190	0.615

Notes: Only 15+ population has been considered. The categories unemployed and students have been excluded in the calculation of the diversity index. HID_1 refers to diversification index calculated on the basis of the main occupation of the household. HID_2 refers to diversification index calculated separately for males, females and all individuals
Source: Field Survey

5.4.3 *Migration for Work*

Because of limited irrigation facilities and declining soil productivity, many households find it very difficult to sustain themselves solely through crop production.[14] Some households, with educated members, see an opportunity for social mobility and better earnings in the urban areas. Aspirations for better education were also an important reason for trying to migrate, particularly among the youth. As a result of all these factors, it has been found that permanent or seasonal migration has emerged as part of the livelihood strategy of households in the recent years (Pani 2017a).

Combining data from all the villages, it is found that nearly 39% of surveyed households has at least one migrant member (Table 5.10). As expected the extent of such households shows some intervillage variations. It ranges from a high of 62% in Devgarh to a low of 14% in Nandpura. So far as the age distribution of migrants is concerned, a substantial majority of them belong to the younger age groups, i.e. 18–25 and 26–33 years. Thus, it is mostly the youth who are migrating out for work. However, it is still part of the household livelihood diversification strategy, because many of them continue to keep in contact with the village and also a substantial majority belong to the category of seasonal or circular migrants. The occupational distribution of migrants, presented in Table 5.11, confirms this. Of all migrants, 45% work as casual manual labour, and 35% work as *Halwai*. Both these types of migrants are seasonal/ circular migrants, who return to the village in particular seasons. Only 20% of the migrants are in salaried jobs in the government and private sector.

[14] These were the most frequently mentioned reasons in conducted FGDs in the study villages.

Table 5.10 Extent of migration and age distribution of migrants

Villages		Percentage of households having at least one migrant member in the last 1 year		Age distribution of migrants				
		No	Yes	18–25Yr	26–33Yr	34–41Yr	42–49Yr	Total
DP Kishroli	N	25	7	0	3	3	1	7
		78.10	21.90	0.00	42.90	42.90	14.30	100.00
Kishroli	N	35	13	5	4	4	0	13
		72.90	27.10	38.50	30.80	30.80	0.00	100.00
Dandoli	N	35	34	13	13	8	0	34
		50.70	49.30	38.20	38.20	23.50	0.00	100.00
Ratan Basai	N	60	58	14	21	17	6	58
		50.80	49.20	24.10	36.20	29.30	10.30	100.00
Rithona	N	34	38	15	12	9	1	37
		47.20	52.80	40.50	32.40	24.30	2.70	100.00
Baghchini	N	10	4	2	4	2	0	8
		71.40	28.60	25.00	50.00	25.00	0.00	100.00
Devgarh	N	5	8	8	8	1	2	19
		38.50	61.50	42.10	42.10	5.30	10.50	100.00
Esah	N	75	19	1	0	0	1	2
		79.80	20.20	50.00	0.00	0.00	50.00	100.00
Nandpura	N	12	2	58	65	44	11	178
		85.70	14.30	32.60	36.50	24.70	6.20	100.00
All	N	291	183	0	3	3	1	7
N		61.40	38.60	0.00	42.90	42.90	14.30	100.00

Source: Field Survey

5.5 Land Degradation

The focus of the study is on land degradation and to see its implications for the development of the area. In this section, data on the experience and perception on land degradation has been presented, from a livelihood perspective. In a rural context where agriculture is the main source of livelihood, land degradation can have a significant impact on household earnings and welfare. In this study, an attempt has been made to examine the impact of land degradation on household-level development through its impact on agriculture.

Agriculture in the region is mostly rainfed. But to a limited extent, irrigation is available through canal and well irrigation. The farmers in the study region use modern agricultural technology, which is demonstrated in the data presented in Table 5.12. However, farmers also consider the formations of ravines and associated

Table 5.11 Occupational distribution of migrants in the sample households

Name of the study villages		Casual manual labour	Halwai	Govt. job	Private job	Total
DP Kishroli	N	1	0	0	0	1
		100.00	0.00	0.00	0.00	100.00
Kishroli	N	1	2	1	3	7
		14.30	28.60	14.30	42.90	100.00
Dandoli	N	15	14	0	2	31
		48.40	45.20	0.00	6.50	100.00
Ratan Basai	N	25	32	1	0	58
		43.10	55.20	1.70	0.00	100.00
Rithona	N	20	6	4	4	34
		58.80	17.60	11.80	11.80	100.00
Baghchini	N	0	0	1	0	1
		0.00	0.00	100.00	0.00	100.00
Devgarh	N	2	0	2	1	5
		40.00	0.00	40.00	20.00	100.00
Esah	N	7	2	4	6	19
		36.80	10.50	21.10	31.60	100.00
Nandpura	N	0	0	2	0	2
		0.00	0.00	100.00	0.00	100.00
All	N	71	56	15	16	158
N		44.90	35.40	9.50	10.10	100.00

Source: Field Survey

Table 5.12 Nature of input use in agriculture

Inputs	Response	No of farmers	Percentage
Use of pump set	Yes	439	92.6
	No	35	7.4
Use of tractor	Yes	466	98.3
	No	8	1.7
Use of chemical Fertiliser	Yes	473	99.8
	No	1	0.2
Use of pesticides	Yes	274	57.8
	No	200	42.2

Source: Field Survey

processes,[15] which is the main form of land degradation, as a significant problem in the region.

Nearly 97% of farmers reported a decline in land productivity in the last 5 years. 64% of farmers believe that the decline in soil productivity is the main reason behind the decrease in crop productivity, while 17% blame it on inadequate rainfall, and

[15] Gully formations, due to gully head erosion in agricultural land and soil of agricultural land and soil is the major concern for the people and farmers.

Table 5.13 Perception about land degradation and its impact

Variables	Response	Number of observations	Percentage
Productivity/soil Quality decreased in the past 5 years	Yes	459	96.8
	No	12	2.5
	Can't say	3	0.6
Main reason for decrease in productivity	Lack of rainfall	78	16.5
	Declining soil fertility	303	63.9
	Availability of water in canal	93	19.6
	Non-response	14	3
Decline in income from agriculture	Yes	452	95.4
	No	21	4.4
	Can't say	1	0.2
Forced migration due to land degradation	Yes	15	3.2
	No	459	96.8
Increased conflict over land	Yes	448	94.5
	No	26	5.5
House affected by *behad*	Yes	18	3.8
	No	402	84.8
	Other related problems	54	11.4
Shortage of firewood	Yes	467	98.5
	No	7	1.5
Total no. of observations		474	100

Source: Field Survey

20% think that inadequate supply of canal water is the main reason (Table 5.13). Probably as a result of the decline in agricultural productivity, 95% of farmers reported a decline in their income. However, only 3.2% reported that they have been forced to migrate because of land degradation.[16] Nearly 93.9% of farmers reported sandiness of the soil as a problem affecting cultivation. A substantial majority felt that the levelling of ravines (Pani 2017a) has contributed to increasing conflicts and decline in firewood availability. The decline in firewood availability affects household welfare significantly, because nearly 84% of the surveyed households use firewood and cow dung cakes as a source energy for cooking (Table 5.14).

The severity of soil loss for agriculture can be judged from the fact that 93% report that soil erosion and runoff are problems during the rains and 84% of them report that it is widely prevalent (Table 5.15).

[16] Damage of their residence and to protect the agricultural land it is very common to shift their houses in nearby agricultural land. Sometimes entire family or any one or two members of family force to shift this information has been shared during FGD and found in earlier study (Pani and Mohapatra 2001, Pani and Carling 2013, Pani 2017a).

Table 5.14 Sources of fuel

Sources of energy for Cooking	No	Percentage
Firewood	4	0.8
Both firewood and cow dung cakes	394	83.1
Both firewood and kerosene	1	0.2
Both firewood and LPG	75	15.8

Source: Field Survey

Table 5.15 Perception on soil erosion and runoff

Variables	Response	No. of observations	Percentage to total no. of observations
Soil erosion and runoff during rain	Yes	368	93.4
	No	26	6.6
Soil erosion prevalence	Widely prevalent	330	83.8
	Moderately prevalent	31	7.9
	Non-responsive	33	8.4
	N	394	100

Source: Field Survey

5.5.1 *Perceptions on Land Degradation Across Farm Sizes*

Rural economy is mainly agriculture dependent, but often it is found that cultivators belonging to different categories experience different kinds of constraints, risks and opportunities. There are suggestions in the literature that small and marginal farmers often find it difficult to borrow from formal credit markets. The literature on risks and uncertainties in agriculture distinguishes between covariant and non-covariant risks. In the case of covariant risks, risks are highly correlated, and hence all farmers are equally affected simultaneously. In the case of non-covariant risks, however, there is heterogeneity in the risk patterns, and some are affected more than the others. Natural hazards like drought, floods and cyclones are examples of covariant risks, where farmers in a particular geographic zone are affected simultaneously. In case of other kinds of environmental hazards, such as land degradation, there is no a priori expectations that all farmers will be equally affected. However, in the case of severe land degradation, all the farmers are likely to be affected simultaneously. The Chambal Badlands are known as a severely affected region. The sampling design followed in this study also ensured that farmers affected by land degradation in a particular geographical zone were selected in the sample. Thus, there is a great deal of similarity in the experience of land degradation by different categories of farmers in the study villages.

It is found that as high as 55% of the farmers have reported some decline in the agricultural land owned by them due to ravines (Table 5.15a). The severity of the problem can be judged from it. The share of farmers reporting such decline in area goes on increasing from marginal to medium-size classes and then decline for the

Table 5.15a Decline in size of land-holdings because of ravines

Size class of land-holdings (in bigha)		Decline in size of land-holdings because of ravines		
		Yes	No	Total
Landless (0.0)	N	1	5	6
	%	16.70%	83.30%	100.00%
Marginal (0.1–2)	N	26	37	63
	%	41.30%	58.70%	100.00%
Small (2.1–6)	N	110	91	201
	%	54.70%	45.30%	100.00%
Semi-medium (6.1–10)	N	76	48	124
	%	61.30%	38.70%	100.00%
Medium (10.1–14)	N	30	7	37
	%	81.10%	18.90%	100.00%
Large (above 14.1)	N	16	27	43
	%	37.20%	62.80%	100.00%
Total	N	259	215	474
	%	54.60%	45.40%	100.00%

Source: Field Survey

larger-size class of holdings. Farmers with medium-sized land-holdings report the highest incidence of such decline in terms of the share of farmers reporting such a decline because of ravines.

Compared to loss of agricultural land, decline in agricultural productivity seems to be a more widespread problem for the farmers. As high as 97% of farmers covered in the survey reported a decline in agricultural productivity. However, this result must be seen in the context of the sampling design of the study. Since farmers were selected on the basis of land in a ravine-affected zone, the result is hardly surprising. As expected, there is not much differences across the farm sizes – an overwhelming majority of farmers report a decline in agricultural productivity in each of the size classes (Table 5.15b). A similar picture emerges with respect to decline in income from agriculture. Nearly 95% of farmers – across the farm sizes – reported a decline in income from agriculture (Table 5.15c).

The process of land loss to *bihads* has very specific characteristics. Often, it is a recurring phenomenon, despite attempts to stop erosion. Thus, boundaries get eroded often. There are spill-over effects of gully formation in particular patches of land. Thus, the action or inaction of particular land owner leads to externalities effects for others. Sometimes changes in agricultural land, such as formation or reappearance of gullies, lead to changes in access to other land. Because of all these reasons, it is found that land-related conflicts are on the rise. Apart from the marginal land owners, other size classes report that more than 95% of farmers in each of the land owner categories feel that land-related conflicts have increased in the last 5 years (Table 5.15d).

Table 5.15b Decline in productivity of crops as a result of land degradation

Size class of land-holdings (in bigha)		Decline in productivity of crops as a result of land degradation in the last 5 years			
		Yes	No	Cannot say	Total
Landless (0.0)	N	4	2	0	6
	%	66.70%	33.30%	0.00%	100.00%
Marginal (0.1–2)	N	60	3	0	63
	%	95.20%	4.80%	0.00%	100.00%
Small (2.1–6)	N	196	4	1	201
	%	97.50%	2.00%	0.50%	100.00%
Semi-medium (6.1–10)	N	122	1	1	124
	%	98.40%	0.80%	0.80%	100.00%
Medium (10.1–14)	N	35	1	1	37
	%	94.60%	2.70%	2.70%	100.00%
Large (above 14.1)	N	42	1	0	43
	%	97.70%	2.30%	0.00%	100.00%
Total	N	459	12	3	474
	%	96.80%	2.50%	0.60%	100.00%

Source: Field Survey

Table 5.15c Decline in income from agriculture

Size class of land-holdings (in bigha)		Decline in income from agriculture			
		Yes	No	Cannot say	Total
Landless (0.0)	N	4	2	0	6
	%	66.70%	33.30%	0.00%	100.00%
Marginal (0.1–2)	N	56	7	0	63
	%	88.90%	11.10%	0.00%	100.00%
Small (2.1–6)	N	193	7	1	201
	%	96.00%	3.50%	0.50%	100.00%
Semi-medium (6.1–10)	N	121	3	0	124
	%	97.60%	2.40%	0.00%	100.00%
Medium (10.1–14)	N	36	1	0	37
	%	97.30%	2.70%	0.00%	100.00%
Large (above 14.1)	N	42	1	0	43
	%	97.70%	2.30%	0.00%	100.00%
Total	N	452	21	1	474
	%	95.40%	4.40%	0.20%	100.00%

Source: Field Survey

The evidence presented here suggest that farmers of all size classes are affected by the problem of land degradation. The problem is so severe that irrespective of their size of holdings, farmers are adversely affected.

Table 5.15d Increase in land-related conflicts

Size class of land-holdings (in bigha)		Increased conflict over land over the last 5 years		
		Yes	No	Total
Landless (0.0)	N	4	2	6
	%	66.70%	33.30%	100.00%
Marginal (0.1–2)	N	56	7	63
	%	88.90%	11.10%	100.00%
Small (2.1–6)	N	192	9	201
	%	95.50%	4.50%	100.00%
Semi-medium (6.1–10)	N	119	5	124
	%	96.00%	4.00%	100.00%
Medium (10.1–14)	N	35	2	37
	%	94.60%	5.40%	100.00%
Large (above 14.1)	N	42	1	43
	%	97.70%	2.30%	100.00%
Total	N	448	26	474
	%	94.50%	5.50%	100.00%

Source: Field Survey

5.6 Impact of Ravines on Crop Farming

The formation of ravines affect agriculture in a number of ways. Primarily the impact of ravine formation on agriculture is manifested through decline in soil productivity. Loss of topsoil during the rainy season impacts the productivity of the soil. The gully head erosion or encroachment of gullies eats into agricultural or unaffected lands. Further, gully erosion makes farming impossible if it is severe.

The farmers try to respond to the severe forms of land degradation in different types. One of the many different ways is, for example, they cultivate crops that require relatively less moisture. Nearly 50% of the farmers in the sample reported mustard as the crop which is more suitable for the degraded land or levelled land[17] (Table 5.16). However, when degraded land are brought under cultivation, the productivity of the land goes down. The average productivity of most of the crops declines when the crops are cultivated in the ravine-affected land (Table 5.17).

[17] Most of the farmers cultivating in levelled land informed that the first crop after levelling is high in productivity but declined steadily over the years.

Table 5.16 Crops if cultivated in less fertile land

Crop	Number	Percentage
Wheat	8	3.9
Bajra	4	2.0
Mustard	102	50.0
Soan	11	5.4
Non-response	79	38.7
Total	204	100

Source: Field Survey

Table 5.17 Yield of crops in ravine-affected land

Name of the crops	No. of observations	Production per bigha (in quintals) in degraded land	Production per bigha (in quintals) in all types of land
Wheat	9	2.12	7.35
Bajra	19	0.78	4.55
Mustard	102	4.87*	6.78
Soan	12	1.78	2.18

Source: Field Survey
Note: Figures in the last column are from Table 6.8
* Bigha is the local unit of measurement

5.7 Coping Strategies

Farmers adopt a number of coping mechanisms to cope up with and mitigate the impacts of land degradation. These include agriculture and land-specific coping mechanisms and those which are outside agriculture. Among the standard responses to gully formation in the Chambal region are to build earth bunds, check dams, gully diversion, ravine land levelling, terrace levelling, bamboo plantation, fitting pipes (Plates 5.1a and 5.1b) in the farmland and road to drain extra accumulated rainwater and reduce soil erosion. Use of *andi* plant (castor oil plant) is also a common practice to minimise gully erosion (Pani 2017a, b). The coping mechanisms adopted by the respondents have been presented in Table 5.18. Building of earthen mounds (Plate 5.2) to stop gully flow or to minimise the flow has been adopted by 73% of farmers. Most of these bunds have been built with individual effort of the farmers. Nearly 74% of farmers reported planting of tree as a coping mechanism against gully erosion. However, 80% of farmers reported that they have opted for levelling of ravine land as a coping mechanism.

In order to understand the coping mechanisms adopted by the farmers in greater detail, we have presented details regarding some of these mechanisms in the subsequent analysis. The information on cost of building bunds have been collected from nearly 266 households. It is found that on an average, it varies from Rs. 6821 to Rs. 2190. Taking all samples together, it is found that the average cost of constructing bunds is around Rs. 3235 (Table 5.19). However, it could be an underestimation because farmers typically do not count their own labour as a component of such

Table 5.18 Coping mechanisms adopted

	Coping mechanisms	No. of farmers	Percentage adopted
1	Tried to stop gully flow by using earthen mound	288	73.1
	1.a With individual effort	255	64.7
	1.b Collective effort with other farmers	7	1.8
2.	Planting trees	291	73.9
3.	Left land fallow	51	14.1
4.	Steep slope cultivation	264	67
5.	Land levelling	378	79.70
6.	Outmigration	15	3.2

Source: Field Survey

Table 5.19 Average cost to build bunds in study villages

Villages	No of observations	Average cost of building bunds (in Rs.)
Dandoli	43	2190.7
Ratan Basai	93	3496.77
Rithona	44	2190.91
Baghchini	7	4428.57
Devgarh	5	3460
Esah	60	3346.67
Nandpura	14	6821.43
Total no. of observations	266	3234.59

Source: Field Survey

costs. Further, the extent of cost to farmers can be judged from the durability of the bunds that are constructed. Nearly 37% of farmers considered bunds to be lasting for only 1 year, while 92.2% reported the bunds to be lasting for less than 2 years (Tables 5.20 and 5.21) (Plate 5.3). In order to cope up with the crop loss and land hunger, farmers in the region opt to cultivate some crops in steep slopes. The important crops and crop combinations that have been cultivated in the steep slopes have been presented in Table 5.22. It is noticed that mustard is the most important crop that is planted on the steep slopes (Plates 5.4a and 5.4b).

Coping mechanisms, to be effective, need an institutional environment. For example, if there are costs and benefits that are not restricted to an individual, an institutional mechanism is needed to resolve the incentive problem. The major institutional forms that can be thought of in this context are state institutions, cooperative institutions of the villagers themselves and non-governmental organisations. The experience and perceptions of the farmers in this regard have been presented in Table 5.22. It is noticed that only 17% of farmers reported the involvements of the panchayats in countering land degradation, whereas 82% reported no involvements of the panchayats. Cooperation among the farmers comes out to be the most important form of intervention. Further, such horizontal cooperation of farmers is found

Table 5.20 Sustainability of bunds in study villages

Sustainability of bunds	Number	Percentage
1 year	108	36.6
2 years	164	55.6
3 years	20	6.8
4 years	2	0.7
8 years	1	0.3
Total	295	100.0

Source: Field Survey

Table 5.21 Crop cultivation in slopes

Crops	Number	Percentage
Wheat	25	6.3
Bajra	16	4.1
Mustard	220	55.8
Soan	10	2.5
Both wheat and bajra	1	0.3
Both bajra and mustard	4	1
Both mustard and soan	7	1.8
Wheat, bajra and mustard	3	0.8
Non-response or did not plant any crop	108	27.4
	394	100

Source: Field Survey

to be more prevalent in the cases of terracing and building of earthen dams. An overwhelming percentage of farmers think that such measures have been effective.

5.8 Land Levelling: Extent, Patterns and Implications

Globally, reclamation of land has been one of the coping mechanisms to counter land degradation. It is an important measure because soil is a non-renewable natural resource. But the way through which land is reclaimed has important implications for the sustainability of land reclamation (Pani 2017a, b). It also has important socio-economic consequences (Pani 2016, 2017a, b). Land levelling has been one of the important ways through which ravine land has been reclaimed in the study region.

The data presented in Table 5.23 indicates that more than 87% of the total sample households have reported some ravine-affected agricultural land. Among the villages, Ratan Basai and Esah have high proportion of ravine-affected households. However, only 2.1% of the total households in all villages have abandoned some land in the last 10 years. And around 80% of the households have tried to level the

Table 5.22 Institutional aspects of measures taken to stop land degradation

Variables	Response	N	Percentage
Initiative by panchayat to stop land degradation	Yes	66	16.8
	No	322	81.7
	Non-response	6	1.5
Steps taken to stop land degradation with other villagers	Yes	365	92.6
	No	29	7.4
Community participation in soil conservation	Contour ridges	2	0.5
	Tree and grass planting	6	1.5
	Manure/crop residues application	1	0.3
	Check dam	1	0.3
	Earthen dam	120	30.5
	Gully plugging	3	0.8
	Gully flow change	11	2.8
	Terracing	90	22.8
	Non-response	160	40.6
Measures taken were successful or not	Yes	362	91.9
	No	7	1.8
	Non-response	25	6.3
	Total	394	100

Source: Field Survey

land. Land levelling is relatively high in Ratan Basai, Dandoli, Nandpura and Baghchini.

Land reclamation is a cost-intensive activity. In the recent years, with the use of heavy machinery like earth movers and tractors in land levelling, the cost of levelling land has increased manifold. In few years back, it used to be done through manual labour using rudimentary tools and techniques. While the better-off households could mobile wage labour for levelling land, the small and marginal farmers were dependent on household labour and exchange labour of friends, relatives and other villagers. The rising per unit cost of levelling land has changed this scenario. Now all kinds of farmers are dependent on heavy machinery for levelling land. Those who cannot afford such expenditures either borrow money from others or forego land levelling. The scale of land levelling has increased as well because of mechanisation (Pani 2017b). The average size of land levelling reported by sample households is 2.5 bigha, but FGDs with affected villagers indicate that people with better resources and connections in urban areas go for land levelling at a much bigger scale. Among the study villages, average size of land levelled is the highest in Kishroli village. In this village, the per household and per bigha expenditure on land levelling are highest as well. The average expenditure per household for land levelling was found to be Rs. 27,902, and average expenditure per bigha of land levelled was Rs. 11,155 (Table 5.24). Given the high cost of land levelling, and the frequency at which gully erosion is found in levelled land, this cost has acquired the

Table 5.23 Land loss and reclamation

Villages	No. and %	Land affected by ravines		Abandoned land in the last 10 years		Tried to level land in the last 10 years	
		Yes	No	Yes	No	Yes	No
DP Kishroli	N	20	12	1	31	15	17
	%	62.50	37.50	3.10	96.90	46.90	53.10
Kishroli	N	43	5	1	47	38	10
	%	89.60	10.40	2.10	97.90	79.20	20.80
Dandoli	N	53	16	3	66	64	5
	%	76.80	23.20	4.30	95.70	92.80	7.20
Ratan Basai	N	116	2	2	116	116	2
	%	98.30	1.70	1.70	98.30	98.30	1.70
Rithona	N	61	11	1	71	55	17
	%	84.70	15.30	1.40	98.60	76.40	23.60
Baghchini	N	12	2	0	14	12	2
	%	85.70	14.30	0.00	100.00	85.70	14.30
Devgarh	N	12	1	0	13	11	2
	%	92.30	7.70	0.00	100.00	84.60	15.40
Easha	N	84	10	2	92	54	40
	%	89.40	10.60	2.10	97.90	57.40	42.60
Nandpura	N	14	0	0	14	13	1
	%	100.00	0.00	0.00	100.00	92.90	7.10
Total	N	415	59	10	464	378	96
	%	87.60	12.40	2.10	97.90	79.70	20.30

Source: Field Survey

nature of a recurring cost for agricultural households. It particularly emerged as a region of indebtedness for the small and marginal farmers (Table 5.25).

An important aspect of levelled land is its legal status. Land reclamation can be done in privately owned land that has been affected by the ravines, but more often than not, it is not confined to legally owned land. In many villages, land levelling has emerged as a mechanism for privatisation of commons (Pani 2017b). Thus, 76% of those who have levelled land have no legal entitlement over the land. However, in villages such as Kishroli, Dandoli, Ratan Basai, Rithona and Devgarh, a relatively higher percentage of households have reported having no *patta* on the levelled land (Table 5.26).

So far as the crops cultivated in the levelled land are concerned, it is found that nearly 75% of farmers cultivated mustard and 13% and 10% cultivated wheat and bajra in the levelled land (Table 5.27). This pattern is also repeated when asked which crops are grown more in the levelled land. The productivity levels show that levelled lands are less productive than other lands. However, it is important to note that the productivity of levelled land might vary over a period of time, depending on changes in soil quality. Nearly 89% of farmers did not agree that levelled land needs

Table 5.24 Village-wise average expenditure on land levelling

Villages	N	Average size of land levelling (in bigha)	Mean cost of land levelling per household* (in Rs.)	Cost of land levelling per bigha (Rs.)**
DP Kishroli	15	2.8	19366.67	6916.67
Kishroli	38	3.8684	57881.58	14962.59
Dandoli	64	2.125	20929.69	9849.26
Ratan Basai	116	2.3103	31495.69	13632.46
Rithona	55	2.0	17909.09	8954.55
Baghchini	12	2.3333	7708.33	3303.57
Devgarh	11	2.9091	20181.82	6937.50
Esah	54	2.6111	25935.19	9932.62
Nandpura	13	3.1923	28000.00	8771.08
All	378	2.5013	27902.12	11154.94

Source: Field Survey

Note: (i) *Excluding those who have not levelled the land; (ii) **Cost as reported by the households

Table 5.25 Legal status of levelled land

		Households having *patta* for levelled land		
Villages	N and %	Yes	No	Can't say
DP Kishroli	N	5	27	0
	%	15.60	84.40	0.00
Kishroli	N	12	35	1
	%	25.00	72.90	2.10
Dandoli	N	32	37	0
	%	46.40	53.60	0.00
Ratan Basai	N	17	101	0
	%	14.40	85.60	0.00
Rithona	N	12	59	1
	%	16.70	81.90	1.40
Baghchinni	N	5	9	0
	%	35.70	64.30	0.00
Devgarh	N	4	9	0
	%	30.80	69.20	0.00
Esah	N	17	76	1
	%	18.10	80.90	1.10
Nandpura	N	6	8	0
	%	42.90	57.10	0.00
Total	N	110	361	3
	%	23.20	76.20	0.60

Source: Field Survey

Table 5.26 Perceptions on profitability of land levelling

Villages	N and %	Is land levelling profitable?	
		Yes	No
DP Kishroli	N	13	19
	%	40.60	59.40
Kishroli	N	35	13
	%	72.90	27.10
Dandoli	N	62	7
	%	89.90	10.10
Ratan Basai	N	106	12
	%	89.80	10.20
Rithona	N	53	19
	%	73.60	26.40
Baghchinni	N	8	6
	%	57.10	42.90
Devgarh	N	11	2
	%	84.60	15.40
Esah	N	49	45
	%	52.10	47.90
Nandpura	N	10	4
	%	71.40	28.60
Total	N	347	127
	%	73.20	26.80

Source: Field Survey

Table 5.27 Cropping pattern and productivity in levelled land

Name of the crops in levelled land	No. of HHs	Percentage of HHs cultivated in levelled land	Production per bigha (Qtl per bigha)	Crops grown more in levelled land	
Wheat	50	13.3	2.70	26	6.9
Bajra	41	10.9	–	30	8
Mustard	283	75.1	1.5	305	80.9
Soan	3	0.8	–	11	2.9
Non-response	–	–		5	1.3
Total no. of observations	377	100		377	100
HHs not cultivated anything	97	20.5		97	20.5

Source: Field Survey

more fertiliser compared to normal land, However, it should be noted that reluctance of farmers to use fertiliser in levelled land could be because of low return to investment from levelled land also, 80% have reported that it requires more labour input than normal land (Table 5.28).

The levelled land is often subjected to further gully erosion. The loose soil gets further eroded in the next monsoon. This problem is considered to be a problem

Table 5.28 Perception regarding inputs required in levelled land

Perception regarding inputs	Need more than normal land	Number	Percentage
Levelled land needs more fertilizer	Yes	48	10.1
	No	420	88.6
	Can't say	6	1.3
Levelled land needs more labour	Yes	379	80.0
	No	95	20.0
Total no. of observations		474	100

Source: Field Survey

Table 5.29 Soil erosion on levelled land

Variables	Response/non-response	No. of observations	Percentage
Soil erosion in levelled land*	Yes	315	66.5
	Yes and extreme	18	3.8
	No	92	19.4
	Cannot say	49	10.3
Katab in agricultural land**			
	Yes	37	7.8
	Yes, and extreme	306	64.6
	No	80	16.9
	Cannot say	51	10.8
Total no. of observations		474	100

Source: Field Survey
Note:: *Soil erosion includes loss of topsoil.**Katab is formation of gullies in levelled land

faced by 70% of the respondents, and *Katab* or gully formation in levelled agricultural land is considered to be a serious problem by 65% of the farmers. In total, 72.4% of farmers have faced such problems in their levelled agricultural land (Table 5.29).

Land levelling is known to have created various social problems. The reason of these conflicts could be varied. Some of these reasons are related with ill-defined property rights over land. But some other conflicts arise due to externality problems. Land levelling by one farmer might create problems for others, such as loss of grazing or forest land, denial of passage, restrictions on movement of livestock, increased gully erosion, soil loss, siltation in nearby water bodies, etc. Apart from the loss of natural resources and reduced natural habitat for the wildlife, the human-animal conflicts are more frequent nowadays. Land levelling or encroachment of ravine land for private use is the major reason for loss of local trees like *tenti* (wild berry), parbati babul, neem, kher, etc., which were in many ways among the alternate sources of livelihoods for the poor. Not having access to the forest land, people are forced to use the crop residue for cooking, which is the cause of respiratory health

Table 5.30 Conflict and other related problems of land levelling

Types of problems	No. of observations	Percentage
Conflict among the households	9	1.9
No conflicts	422	89.0
Problems like boundary creation	20	4.2
Money and labour problems	21	4.4
Irrigation and water holding problem	2	0.4
Total no. of observations	474	100

Source: Field Survey

problems for the women and children of the region[18] (Pani 2017b). However, most of the farmers in our surveys denied the presence of any such conflicts (Table 5.30).

5.9 Policy Approaches Towards Controlling Land Degradation and Its Problems

There are several programmes and policies adopted to deal with the problems of land degradation in the study region. India has a long history of degraded land reclamation with support from various government agencies. Combating degradation include reforestation and levelling of ravine and gullied land and reclamation of acidic, saline and waterlogged soils.

Many attempts had been taken to reclaim the land in different phases since post-independence. The Madhya Pradesh state government had introduced few projects in the 1950s (Verma et al. 2012; Pani 2016) including in sites such as Chonda, Bagchini and Nayakpura in Morena district. The central government sanctioned a major pilot project in 1955, under the Central Tractor Organization (CTO) in the Nayakpura area, which used machinery to reclaim a land of 400 ha.[19] Around 142 ha had been reclaimed for agricultural use by the CTO in 1959 under the same project. Government of India and the Ministry of Home Affairs in the year 1972 aimed at the reclamation of over 55,000 ha of Chambal Badlands for agriculture in the states of Madhya Pradesh, Uttar Pradesh and Rajasthan (Haigh 1984) in a 7-year plan for Ravine Reclamation. The programme was not successful in preventing the soil erosion of the area (Sebastian 2001).

In 1967, the Government of India constituted a Central Ravine Reclamation Board aiming at systematic ravine control and development of the region. The National Policy adopted had identified two broad objectives:

[18] Some of the young women in the villages of Esah and Bagchini have shared their concern regarding health problems they are facing these days, during the survey.

[19] Based on interview with field-level staff of the Department of Agriculture, GoMP

'(a) to arrest further growth of the ravine land and (b) to utilise the already reclaimed land for productive purposes ... depending upon land capability and the practical consideration of social and economic conditions. The following steps were envisaged for ravine reclamation: (i) survey and planning; (ii) ravine classification and use; (iii) land rights and transfers; (iv) social problems; (v) operational research and evaluation' (Chambal Command Area Development, Authority undated, 22).

'In 1971, the Indian Government envisaged a large-scale development project for the Chambal Valley, with an estimated budget of Rs 1224 crores to be spent in the next 27 years. The four-phased project was aimed at developing 55,000 ha of land for agriculture and 27,500 ha for forestry and pastureland, in the first phase in three different states. However, the project did not take off' (Sebastian 2001). However, only 2137 ha of ravines was reclaimed by around 18 projects over 25 years of time (Sebastian 2001), and only few of them achieved the target.

According to a report by the Planning Commission, 'The two main categories of degraded lands by ownership – private and government – received different treatments in official programmes in the past. Whereas soil and moisture conservation measures were attempted by the Agriculture Departments on private lands, social forestry plantations were undertaken on government wastes by the Forest Departments of the state governments. These two programmes suffered from two common weaknesses: First, there was no integrated land management – the two programmes ran in isolation. Second, these were, till very recently, entirely departmental affairs with no participation from the people' (Planning Commission, undated). The National Commission on Agriculture (para 17.2.29 part V) noted: 'The economic benefit should not be the sole consideration in the reclamation of ravines'. Citing poverty and the disturbed law and order situation in the region, it argued that: '[a]t present hundreds of habitations and their valuable tablelands are threatened with marching ravines which ravines which are taking a toll of about 800 hectares of valuable agricultural land every year costing nearly Rs 41053 crores. In view of the above, ravine reclamation should receive national priority and investment should not be denied on account of narrow or unfavourable benefit–cost ratio' (Government of India 1976). Thus, levelling of ravine land, despite the huge costs involved, was thought of as a necessary intervention in the study area.

The Madhya Pradesh Government also had initiated a plan for levelling the ravines using bulldozers, which did not achieve much success. Reforestation through aerial seeding in the ravines was also part of a project in the 1980s, to develop 12,009 hectares of forest every year (Sebastian 2001; Mudgal 2005). Despite several attempts to level ravines, the plans have not been successful. Efforts by NGOs, though successful at the micro-level, were undermined because of ownership issues, as most of the land was categorised as government land (Pani 2017a).

Thus, most of them are reclaiming the ravine land on their own individual initiatives. But there is always uncertainty and stress on land ownership as most of the ravine land either comes under common property resources or forest land. *Pattas* or legal entitlement over reclaimed land is given by the government to farmers belonging to socially deprived and landless categories (Sebastian 2001). But FGDs suggest that many of them were not able to hold on to these lands for long, partly because of the heavy cost associated with the management of reclaimed land. Lack of people participation and uncertainty of ownership for reclaimed land as reasons for unsuccessful ravine-reclamation projects, which were supported by the earlier researchers too, were criticised (Verma et al. (2012) and Yadav and Bhushan (2002)).

In recent years, the Government of Madhya Pradesh has included in its Agenda 2018 'programmes specific to developing ravines, in the region' and 'a intensive strategy to transform the ravine lands to an agricultural land' (Department of Planning, Economics and Statistics (DPES) 2013). According to recent press reports (Trivedi 2016), the GoMP has applied for financial assistance of Rs 1200 crore from the World Bank to reclaim the ravine lands, including a plan to level 18,000 ha of ravine lands.

The recent approach towards land reclamation has been conceived within the framework of watershed development, particularly under the National Watershed Development Project for Rainfed Area sponsored by the central government. During the eighth 5-year plan in 1990–1991, in 385 blocks of Madhya Pradesh, having less than 30% assured irrigated area, an ambitious watershed development programme was launched in 385 selected watersheds covering 852,753 ha (Pani 2017a). Although the provision of earthworks under the Mahatma Gandhi National Employment Guarantee Act, which guarantees a minimum of 100 days of casual employment per family, has also been used for improvements in agricultural productivity, employment guarantees and so on, their synergy with land reclamation projects should not be taken for granted. Yet it was noticed in few occasions that funds under MGNREGA are being spent to construct check dams and other earthworks related to land reclamation. There appears to have a shift in the policy stance towards the ravines in the recent years. On the one hand, there are not much land-based interventions on the ground specifically aiming at addressing the problem of the ravines. However, there are indications that land reclamation has again been given a serious consideration, although the focus has shifted away from agriculture to land reclamation for industries – a policy that has significance in the context of rising conflicts over land in India (Business Line 2014; Pani 2017a).

On the basis of field survey and discussions with grassroots-level government functionaries, it was noticed that land reclamation initiatives in the study region suffer from several problems related to faulty conceptualisation and poor implementation.[20] First, there is a lack of serious scientific studies and cost-benefit analysis of the specific reclamation strategies in specific micro-contexts. A great deal of ad hoc conceptualisation of land reclamation has been noticed on the ground.[21] According to a study 'the best way to reclaim gullied and ravine cultivated land is to retire the sick lands from active cultivation in order to undertake different gully reclamation measures', but it is highly unlikely that farmers will agree to retire their lands

[20]These findings have been discussed in greater detail in Pani 2016, 2017a, b; Marzolff and Pani 2017a.

[21]On several occasions, farmers expressed dismay over the thrust of the programmes implemented by the government. Similarly, the ground-level staff expressed their lack of comprehension regarding the conceptualisation of the various components of the programmes they were involved in implementing.

because of a lack of alternative livelihoods and also as 'there is no provision of suitable financial compensation to the farmers from any government or nongovernment agency in case of voluntary retirement of gully-infested lands from active cultivation'. Ranga et al. (2016) have argued that levelling practices are manifestations of growing land hunger, and a recent study found that 'levelling activities in badlands could, reasonably, be explained by a fitted model including five explanatory variables viz. distance from the Chambal River, distance from the river channels, distance to settlement, slope and catchment area'.

Secondly, there has been no attempt to involve the local people or their practice-based knowledge to the land reclamation programmes. Several studies show that incorporation of indigenous knowledge helps in local ownership and management of the programmes. It is not necessary that the practices followed by people on the ground are scientific and sustainable in the long run. In fact, on the basis of their own initiatives, 'local farmers are using the interfluves, gully valley sides, and gully channel beds for cropping after sloping and removal of natural vegetal covers such as grass, wild bushes, and dwarf trees. Such cropping has adverse effects on soil erosion, and there has hardly been any attempt to create a framework for internalisation of such externalities' (Pani 2016). The reliance on a top-down approach to land reclamation has resulted in parallel existence of programmes and individual coping strategies. There are hardly any conversation and mutual learning that are taking place on the ground.

Thirdly, the livelihood-based initiatives suffer from a similar mismatch with the requirements of the local environment. Livelihood support programmes, if successful, programmes aiming at support to livelihoods indirectly help in reducing the pressure on fragile environments. However, some 'livelihood-enhancing efforts such as those based on livelihood support have an ambiguous affect, as while more productive and sustainable livelihoods based on animal husbandry would enhance employment and earnings, they might lead to overgrazing and destruction of bihads as well. However, at the implementation level these interfaces are neither being integrated into project formulation nor are being coherently operationalised' (Pani 2016). Fourthly, the lack of horizontal coordination among various state governments involved in land reclamation in the region (i.e. Rajasthan, Madhya Pradesh, and Uttar Pradesh) has limited the scope in scaling-up of successful interventions.

5.10 Conclusion

Land degradation in the form of gully erosion is a serious problem in the study area. Some of its important dimensions have been discussed in this section on the basis of the household-level survey conducted in nine villages of Morena district. The status of the livelihoods of the villagers in the study area and its linkages with land degradation were analysed in this chapter. While agriculture is the main source of livelihood for a majority of farmers, in terms of share of income, non-farm employment has started to play a major role. Along with participation in the local, rural non-farm economy, some villagers, particularly those belonging to the younger age groups, have started to migrate to urban areas as casual workers in the urban informal sector.

The problem of land degradation is faced acutely by most of the agriculture-dependent population. The villagers have started to adopt various coping mechanisms to overcome the problems of gully erosion. These include construction of bunds, cropping pattern changes, planting of trees and cultivation on slopes. These coping mechanisms have important implications for cost of production. The process of land levelling was also investigated through this survey. The results show the significance of land levelling as well as various problems associated with it. Thus, while farmers have been trying to cope up with the challenges of gully erosion in this region, there is yet to be a sustainable and equitable solution to their problems associated with land degradation.

References

Business Line (2014) Madhya Pradesh keen to turn chambal valley into industrial area. The Hindu Business Line, Mumbai, 26 July. http://www.thehindubusinessline.com/news/national/madhya-pradesh-keen-to-turn-chambal-valley-into-industrial-area/article6254758.ece. Accessed 22 February 2016

Department of Planning, Economics and Statistics (DPES) (2013) Madhya Pradesh Vision 2018. An Agenda for Development, Change and Good Governance. Bhopal. Government of Madhya Pradesh. http://mpplanningcommission.gov.in/Vision2018English.pdf [27 July 2017]

Ellis S, Taylor D, Masood KR (1993) Land degradation in Northern Pakistan. Geography 78(1):84–87. Retrieved from http://www.jstor.org/stable/40572233

Government of India (1976) National Commission on agriculture report, 1976. Ministry of Agriculture and Irrigation, New Delhi

Haigh MJ (1984) Ravine erosion and reclamation in India. Geoforum 15(4):543–561

Marzolff I, Pani P (2017) Dynamics and patterns of land levelling for agricultural reclamation of erosional badlands in Chambal Valley (Madhya Pradesh, India). Earth Surf Process Landf. https://doi.org/10.1002/esp.4266

Mudgal MK (2005) Socio-economic impact of ravine lands: a case study of river Chambal Basin of state of Madhya Pradesh, India. Geophys Res Abstr 7:005291–005296

Pani P (2016) Controlling gully erosion: an analysis of land reclamation processes in Chambal Valley, India. Dev Pract 26(8):1047–1059

Pani P (2017a) Ravine erosion and livelihoods in semi-arid India: implications for socioeconomic development. J Asian Afr Stud 53(3):437–454

Pani P (2017b) Chambal without ravines. Down Earth 26(8):80–81

Pani P, Carling P (2013) Land degradation and spatial vulnerabilities: a study of inter-village differences in Chambal Valley, India. Asian Geogr 30(1):65–79. https://doi.org/10.1080/10225706.2012.754775. Routledge

Pani P, Mohapatra SN (2001) Delineation and monitoring of gullied and ravinous lands in a part of lower Chambal Valley, India, using remote sensing and GIS. 22nd Asian conference on remote sensing, vol 54, pp 5–9

Ranga V, Poesen J, Van Rompaey A et al (2016) Detection and analysis of badlands dynamics in the Chambal River valley (India), during the last 40 (1971–2010) years. Environ Earth Sci 75(183):1–12

Sebastian S (2001) Ravine reclamation projects fruitless. The Hindu (17 June)

Trivedi S (2016) MP seeks Rs 1,200 cr World Bank loan for ravine rec- lamation. Business Standard (17 November)

Verma GP, Singh YP, Dubey SK (2012) Watershed based reclamation and control of Chambal ravines. In: Dubey SK, Dubey RK, Singh AK, Panda PK, Kala S, Sharda VN (eds) Conservation of natural resources for food and environmental security. Satish Serial Publishing House, Delhi, pp 329–342

Chapter 6
Conclusion

Abstract The conclusion and summary of the findings have been placed in the last chapter. The overall findings of the study support the assertion that the relationship between land degradation and socio-economic development is complex and is essentially a two-way process. While land degradation is found to have impacted socio-economic development of the region, land degradation itself has also been aggravated by various anthropogenic causes, including changes in the agricultural practices. In the study region, it was observed that the land- and agriculture-based responses of the affected population to land degradation further aggravate land degradation.

Keywords Land degradation · Socio-economic development · Linkages · Development policy

6.1 Introduction

Land degradation is recognised as one of the severe global environmental challenges, having innumerable economic, social, ecological and environmental implications. Combating land degradation has been an important objective of socio-economic development in many regional contexts. To ensure food security for all sections of people, land is counted to be one of the most important natural resources in the world. The economy of the underdeveloped and developing countries is predominantly reliant on land efficiency, as a substantial section of people depend on agriculture or land-related resources for their livelihoods. Land degradation poses a considerable challenge to agricultural growth and poverty reduction in India as well. Therefore, the area with degraded land, in general, has lower level of socio-economic development (Bai et al. 2008). Appropriate institutional actions could combat or lower degradation with the same circumstances of snowballing population pressure (Reddy 2003).

The magnitude and expanses of degradation are noteworthy in India. Being an agrarian economy, any aggravation in the situation would become threats to food security of the nation. With stagnant growth of yield in the traditional green revolu-

P. Pani, *Land Degradation and Socio-Economic Development*, Advances in Asian Human-Environmental Research,
https://doi.org/10.1007/978-3-030-42074-1_6

tion in regions of India due to a combination of factors including environmental degradation (Bhalla and Singh 2012), India's food security situation is critically dependent on the way land degradation is contained in the future. As of now, however, the adverse impacts of degradation are limited to certain areas. As far as mitigating measures are concerned, the ARPU at Ahmedabad has come up with the necessary policy interventions at the regional level (Wadia 1996). Cautious management of all these mechanisms is significant for subsequent development of sustainable land use practices.

Chambal region, the focus of the study, is among the most adversely affected regions in India in terms of land degradation. Livelihoods in the study region, Chambal, are primarily agricultural, and a considerable segment of the rural workforces are reliant on agriculture, either as cultivators or as agricultural labour. Therefore, land is considered to be the most precious asset for the socio-economic development of the region.

6.2 Key Findings of the Study

The study aimed at providing a comprehensive assessment of land degradation as well as its socio-economic implications. It not only attempted to assess the temporal and spatial dynamics of land degradation in the study area but also investigated interlinkages of land degradation with human-induced activities, with a particular focus on agriculture. The coping strategies of people affected by land degradation have also been investigated. In particular, the extent, methods and implications of the reclamation of the ravine lands were among the focus of the study. The responses to land degradation through government policies and programmes were also examined to assess the future prospects for land-based development in this fragile ecological region. Given the objectives of the study, a mixed-method approach was followed integrating geospatial techniques such as remote sensing and GIS with field survey methods involving quantitative and qualitative survey techniques. To understand the implications of land degradation, a household survey was conducted in seven severely ravine-affected villages. The villages were identified for survey through the ravine map generated from satellite images of 2014 and superimposed with the district village boundary map. Through structured questionnaires, various dimensions of land degradation and its implications for the livelihoods of the households were investigated. To substantiate the findings, FGDs were conducted with the villagers and farmers of all sections. In-depth, open-ended interviews were conducted with key informants, farmers, Panchayat Samity members, government officials and social activists from the region.

It has been found that in India, 32.07% (105.48 mha) of the total geographical area of the country was affected by various types of land degradation (ISRO 2007). Out of the total degraded area, 10.21% (33.56 mha) was due to water erosion only. State-level estimates from the same study suggest that Madhya Pradesh has 13.5 mha of degraded land. About 37.92% was due to water erosion only (ISRO

2007). It has been estimated that 3.98 mha of ravines occur along the rivers Yamuna, Chambal, Mahi, Sabarmati and others (Ministry of Agriculture 1984). The Chambal region is infamous for its inaccessible rugged terrain due to this water-borne land degradation problem called ravine, locally known as *behad.* Out of four major ravine zones in India, Chambal is the largest one. Chambal region is among the worst affected region of India due to its nature and degree of degradation. But the nature of and dynamics of degradation need a continuous monitoring and assessment which is possible only using geospatial technology and repetitive field investment. Using both these methods, it has been found the ravine and gullies are severe in nature. The key findings on the spatiotemporal changes in ravine formation in the study area have been listed below.

- Firstly, as per our estimation, the total ravine area from 1082.71 km^2 in 1974 has been reduced to 415.98 km^2.
- Secondly, three distinct types of ravines, deep, moderate and shallow ravine, have been identified through our field survey. All the types of ravine exist together. Through the types of ravines, the degree of ravine erosion can be identified.
- Thirdly, the ravine process is still very active, although because of human-induced factors, such as land levelling, the total ravine area has been reduced in the last 40 years of time.

It has been established globally that land degradation is interlinked with human-induced activities. There are several activities including unscientific agricultural practices on slopes and levelled ravine land that result in a huge amount of irreversible soil loss in every rainy season. There are many off-site and on-site impacts and implications of anthropogenic activities in the study area.[1]

In this study, an attempt has been made to understand the implications of land degradation for agriculture and rural development. Considering that the formations of ravines and associated processes are a significant problem for agricultural productivity and development in the region, the linkages were investigated through a village-level analysis and also on the basis of the quantitative and qualitative insights gathered through field research.[2] Due to active ravine formation processes, there are several implications like social isolation, fragmentation of the household and circular migration, and problem of getting higher education for the girls of the region is a major concern for the area (details in Chap. 5).

The study area is known for its largest share, witnessing the most severe form of land degradation in India. According to the data from the population census, in 2011, nearly 53% of the rural workers were cultivators, and 26% of rural workers were agricultural labourers. Thus, nearly 79% of rural workers were dependent on agriculture, suggesting limited availability of alternative livelihood options.

[1] For a detailed discussion and evidences, see Chaps. 2 and 5.

[2] The village-level analysis has been presented in Chap. 4, and the analysis of household-level primary data has been presented in Chap. 5.

Similarly, in the sample survey, it was found that around 81% of surveyed households were dependent on cultivation as their principal source of income.

Chambal is affected by the ravine erosion which is the cause of severe topsoil loss, and it has tremendous economic implications (Pani and Carling 2013). The implications of gully or ravine erosion are not only limited to the loss of agricultural land; in numerous localities it has also caused shifting of villages either in a village level or in household level (Pani and Mohapatra 2001; Pani 2017b). The ravine and gully network expansion and headward erosion are having major impact on physical and sociological environment of the surrounding region. The loss of topsoil and gully encroachment has resulted in loss of productive capacity of agricultural land and damages of property and roads. The long history of land degradation of this region might have played a crucial role for underdevelopment. Analysing village-level data available through the Census of India (2001), Pani and Carling (2013) have shown that villages located within the degraded areas of Morena District have consistently lower levels of socio-economic development than those located outside the area. The analysis of population census data from the 2011 census, presented in Chaps. 3 and 4 of the book, points to the relative backwardness of the region as a whole. Within this relatively backward region, villages located in the deep ravines have lower literacy rates and lower levels of infrastructure development.

The formation of ravines affects agriculture in a number of ways. Primarily the impact of ravine formation on agriculture is manifested through decline in soil productivity. Loss of topsoil during the rainy season impacts the productivity of the soil. The gully head erosion or encroachment of gullies eats into agricultural or unaffected lands. Further, gully erosion makes farming impossible if it is severe.

A large section of farmers and villagers pointed that levelling of ravines has contributed to increasing conflicts and decline in firewood availability. The decline in firewood availability affects household welfare significantly, because for nearly 84% of surveyed households, firewood and cow dung cakes are the main sources of energy for cooking.

The ecological impact of levelling Badlands has not only resulted in the demolition of unique landscapes, it also created problem in social harmony and increased tension and has led to distress migration. A more holistic approach towards the ravine land can minimise the outcomes of soil erosion. The usefulness of every landscape cannot be confined to its value in economic terms. One of the significant impacts of large-scale levelling is the disappearance of common lands, which include grazing lands. Levelling *bihads* has given an opportunity to wealthy farmers to privatise commons. During field survey, it has been noticed that many landless people, who earlier had access to the *bihads* to graze their cattle, could no longer have the freedom to venture or use the common lands. This has forced to a decline in the livestock population, and the organic links between rainfed agriculture and livestock rearing have collapsed. The region already has a long history of oppression and crime; therefore, this situation led to social inequality and conflicts across the region. Due to this, many landless and marginal farmers are migrating to nearby cities in search of work.

Farmers and locals adopt a different type of coping mechanism to cope up with and mitigate the impacts of land degradation. These include agriculture and land-specific coping mechanisms and those which are outside agriculture. Among the standard responses to gully formation in the Chambal region are to build earth bunds, check dams, gully diversion, ravine land levelling, terrace levelling, contour binding, bamboo plantation, fitting pipes in the farmland to drain extra accumulated rainwater and reduce soil erosion. Use of *andi* plant (castor oil plant) is also a common practice to minimise gully erosion (also see Pani 2017a, b). The details of the coping mechanisms adopted by the respondents have been presented in Table 5.18. Building of earthen mounds to stop gully flow or to minimise the flow has been adopted by 73% of farmers. Most of these bunds have been built with individual effort of the farmers. Nearly 74% of farmers reported planting of tree as a coping mechanism against gully erosion. However, 80% of farmers reported that they have opted for levelling the ravine land as a coping mechanism. To sum up, the field investigations suggest that farmers adopt a range of coping mechanisms, from changes in cropping pattern to construction of dams and levelling land to mitigate the impacts of ravines. However, most of the coping mechanisms are based on individual efforts rather than collective action.

The overall findings of the study support the assertion that the relationship between land degradation and socio-economic development is complex and is essentially a two-way process. While land degradation is found to have impacted socio-economic development of the region, land degradation itself has also been aggravated by various anthropogenic causes, including changes in the agricultural practices. In the study region, it was observed that the land- and agriculture-based responses of the affected population to land degradation further aggravate land degradation. Thus, initially people try to respond to land degradation by processes such as land levelling, which in turn leads to further land degradation. There are specific socio-economic factors affecting land degradation; and any long-term development of the region would require an intervention in terms of creating sustainable land use practices and responses.

6.3 Policy Suggestions

Land is considered to be an important asset, and hence land reclamation has received attention from policy makers since independence. A cursory glance at the major policy initiatives suggests that there were two basic thrusts of the policy approach towards land degradation in the Chambal region. Firstly, land reclamation, mostly through land levelling, has been the major objective of policies towards land in the Chambal region. Secondly, to a lesser extent, reclaimed land was considered to be a kind of surplus land that could be distributed to the weaker sections, such as marginalised social groups and landless labourers. In this study, an attempt was made to understand the effectiveness of the ongoing policy interventions on the basis of qualitative surveys and interviews with farmers and government officials.

Land levelling has critical consequences on soil properties, soil loss, soil structure and soil hydrological processes. On-site land degradation often created extensive off-site effects such as silting of rivers and local water bodies. The sustainability of such newly created lands which are mostly used for agriculture remains a challenge. Apart from the physical aspect of it and the privatisation of the common land, it disturbs the social harmony and environment of the region. To look into this, there is an urgency to bring an appropriate policy to protect the ecology of the area. There are many unsuccessful attempts to arrest the degradation of the area. It is the time now to reinvestigate the failure of all attempted programmes and adopt a scientific social and environment-friendly approach to bring up a new policy for the region.

There is no such exclusive policy introduced for the ravine-affected region in recent years. There are few attempts with the existing ongoing development programmes like MGNRGA. It has been found that these are not structured enough to deal with the problems of this region. During the field investigation, it has been noticed that building check dams in ravine watershed is an effective measure to arrest soil loss and accumulate sediments. It is a very effective measure to store water after rain, but looking at the set of climatic conditions of the region, the stored water cannot be sustained given the high evapotranspiration rate in the long run. Therefore, the purpose of improving groundwater condition and irrigating the farmlands gets defeated. Plantation is one of the major investments of the government and is a matter of concern, but it is not likely to succeed unless villagers are involved in the plantation. There should be a policy to generate resources from the adopted measures to maintain check dams. Effective policy intervention should take into account peoples' perceptions of land degradation and its implications for their livelihoods.

To sum up, the following policy suggestions emerge from the study. Firstly, there should be a comprehensive and integrated ecological policy towards the Chambal region. Its ecological uniqueness and fragility should be recognised by developing locally relevant interventions to address the problems of land degradation. Secondly, there should be intergovernmental coordination among the state governments, preferably involving the district and panchayat level bodies from the entire Chambal region to design and implement various development programmes. An area planning approach involving the government, private sectors and civil society groups should be followed. Thirdly, a bottom-up and multi-stakeholder approach should be adopted to involve people from the region as well as experts and scientists to create an integrated and long-term policy approach towards land in the Chambal region. Fourthly, economic incentives should be built into the land conservation measures, so that sustainable land use practices are encouraged and unscientific and harmful practices are discouraged. Finally, the ownership rights over the entire *behads* should be clearly demarcated, with the help from digital technologies if needed, so that privatisation of commons could be halted.

References

Bai ZG, Dent DL, Olsson L, Schaepman ME (2008) Proxy global assessment of land degradation. Soil Use Manag 24(3):223–234. Wiley Online Library

Bhalla GS, Singh G (2012) Econmomic liberalisation and Indian agriculture- a district-level study. SAGE Publication, New Delhi

Ministry of Agriculture (1984) Report of the working group on reclamation and development of ravines for formulation of the five year plan, New Delhi

Pani P (2017a) Ravine erosion and livelihoods in semi-arid India: implications for socioeconomic development. J Asian Afr Stud 53(3):437–454

Pani P (2017b) Chambal without ravines. Down Earth 26(8):80–81

Pani P, Carling P (2013) Land degradation and spatial vulnerabilities: a study of inter-village differences in Chambal Valley, India. Asian Geogr 30(1):65–79. https://doi.org/10.1080/10225706.2012.754775. Routledge

Pani P, Mohapatra SN (2001) Delineation and monitoring of gullied and ravinous lands in a part of lower Chambal Valley, India, using remote sensing and GIS. 22nd Asian conference on remote sensing, vol 54, pp 5–9

Reddy VR (2003) Land degradation in India: extent, costs and determinants. Econ Pol Wkly 38(44):4700–4713

SAC, ISRO (2007) Desertification & land degradation Atlas of India. Space Application Centre/ISRP, Ahmedabad

Wadia FK (1996) Agro-climatic regional planning at zone level. In: Basu DN, Guha GS (eds) Agro-climatic regional planning in India, Part one: concepts and applications. Concept Publishing Company, New Delhi, pp 85–134

Bibliography

Bai ZG, de Jong R, van Lynden GWJ (2011) An update of GLADA - global assessment of land degradation and improvement - GLADA Report update. (Report 2010/08). Wageningen: ISRIC - World Soil Information

P. Pani, *Land Degradation and Socio-Economic Development*, Advances in Asian Human-Environmental Research,
https://doi.org/10.1007/978-3-030-42074-1

Index

P. Pani, *Land Degradation and Socio-Economic Development*, Advances in Asian Human-Environmental Research,
https://doi.org/10.1007/978-3-030-42074-1